Bridge Deck Behaviour

Twin-cell concrete box-girders of Millbrook Flyover, Southampton, England; designed by Gifford and Partners. Photograph E.C. Hambly.

Bridge Deck Behaviour

E.C. Hambly FEng, FICE
Consulting Engineer
Visiting Professor at the University of Oxford
in the Principles of Engineering Design

CRC Press is an imprint of the
Taylor & Francis Group, an **informa** business

A TAYLOR & FRANCIS BOOK

CRC Press
Taylor & Francis Group
6000 Broken Sound Parkway NW, Suite 300
Boca Raton, FL 33487-2742

First issued in paperback 2019

© 1976, 1991 E.C. Hambly
CRC Press is an imprint of Taylor & Francis Group, an Informa business

No claim to original U.S. Government works

ISBN-13: 978-0-419-17260-4 (hbk)
ISBN-13: 978-0-367-86342-5 (pbk)

Typeset in 11/13pt Times by EJS Chemical Composition, Midsomer Norton, Bath, Avon

This book contains information obtained from authentic and highly regarded sources. Reasonable efforts have been made to publish reliable data and information, but the author and publisher cannot assume responsibility for the validity of all materials or the consequences of their use. The authors and publishers have attempted to trace the copyright holders of all material reproduced in this publication and apologize to copyright holders if permission to publish in this form has not been obtained. If any copyright material has not been acknowledged please write and let us know so we may rectify in any future reprint.

Except as permitted under U.S. Copyright Law, no part of this book may be reprinted, reproduced, transmitted, or utilized in any form by any electronic, mechanical, or other means, now known or hereafter invented, including photocopying, microfilming, and recording, or in any information storage or retrieval system, without written permission from the publishers.

Trademark Notice: Product or corporate names may be trademarks or registered trademarks, and are used only for identification and explanation without intent to infringe.

British Library Cataloguing in Publication Data
Hambly, E. C.
   Bridge deck behaviour. – 2nd ed.
   1. Title
   624.1

Library of Congress Cataloguing in Publication Data
Hambly, Edmund C.
   Bridge deck behaviour/Edmund C. Hambly. – 2nd ed.
   Includes bibliographical references and index.
   1. Bridges—Floors. I. Title
TG325.6.H35   1991
624′.2563—dc20   91–2142

Visit the Taylor & Francis Web site at
http://www.taylorandfrancis.com

and the CRC Press Web site at
http://www.crcpress.com

This book is dedicated to my wife Elizabeth, without whom it would not exist, and to the late Kenneth H. Roscoe and Stuart G. Spickett from whom we both continue to draw inspiration.

Contents

Acknowledgements	xi
Preface	xiii
Notation	xvii

1. Structural forms and calculation methods — 1
  1.1 Introduction — 1
  1.2 Structural forms — 2
  1.3 Safety of methods — 17
  1.4 Hooke's law and Young's modulus — 21
  References — 22

2. Beam decks and frames — 24
  2.1 Introduction — 24
  2.2 Types of beam deck — 24
  2.3 Bending of beams — 26
  2.4 Torsion of beams — 39
  2.5 Computer analysis of continuous beams — 45
  2.6 Construction sequence — 45
  2.7 Frame and arching action — 47
  2.8 Short-term and long-term behaviour — 50
  References — 52

3. Slab decks — 53
  3.1 Introduction — 53
  3.2 Types of structure — 53
  3.3 Structural action — 55
  3.4 Rigorous analysis of distribution of forces — 61
  3.5 Grillage analysis — 61
  3.6 Grillage examples — 68
  3.7 Interpretation of output — 72
  3.8 Moments under concentrated loads — 73

	3.9	Shear-key slab decks	74
	3.10	Grillage analysis of shear-key slab	75
		References	81
4.	**Beam-and-slab decks**	82
	4.1	Introduction	82
	4.2	Types of structure	83
	4.3	Structural action	83
	4.4	Grillage analysis	85
	4.5	Grillage examples	89
	4.6	Application of load	92
	4.7	Interpretation of output	94
	4.8	Torsionless design	95
	4.9	Bracing	99
	4.10	Slab membrane action in beam-and-slab decks	102
		References	104
5.	**Multicellular decks**	106
	5.1	Introduction	106
	5.2	The shear-flexible grillage	106
	5.3	Grillage mesh	107
	5.4	Modes of structural action	108
	5.5	Section properties of grillage members	122
	5.6	Load application	127
	5.7	Interpretation of output	127
	5.8	Comparison with finite strip method	132
		References	133
6.	**Box-girder decks**	135
	6.1	Distortion of single-cell box-girder	135
	6.2	Methods of calculation	137
	6.3	BEF analysis of box-girder	137
	6.4	Space frame analysis of box-girder	141
	6.5	Grillage analysis of box-girder	145
	6.6	Grillage analysis of multiple box-girder deck	149
	6.7	Grillage analysis of a multispan box-girder	151
		References	156
7.	**Space frame methods and slab membrane action**	157
	7.1	Truss space frame	157
	7.2	McHenry lattice	160
	7.3	Cruciform space frame	161
	7.4	Slab membrane action	165
	7.5	Downstand grillage	167

	7.6	Effects of slab membrane action on beam-and-slab deck behaviour	174
		References	176
8.	**Shear lag and edge stiffening**	177
	8.1	Shear lag	177
	8.2	Effective width of flanges	177
	8.3	Edge stiffening of slab decks	182
	8.4	Upstand parapets to beam-and-slab decks	183
	8.5	Service bays in beam-and-slab decks	185
		References	187
9.	**Skew, tapered and curved decks**	188
	9.1	Skew decks	188
	9.2	Tapered decks	192
	9.3	Curved decks	193
		References	198
10.	**Distribution coefficients**	199
	10.1	Introduction	199
	10.2	Some published load distribution charts	200
	10.3	Influence lines for slab, beam-and-slab and cellular decks	205
	10.4	Application of charts to slab deck	208
	10.5	Application of charts to beam-and-slab deck	213
	10.6	Application of charts to cellular deck	216
		References	221
11.	**Temperature and prestress loading**	222
	11.1	Introduction	222
	11.2	Temperature strains and stresses in simply supported span	222
	11.3	Temperature stresses in a continuous deck	227
	11.4	Grillage analysis of temperature moments	229
	11.5	Differential creep and shrinkage	230
	11.6	Prestress axial compression	231
	11.7	Prestress moments due to cable eccentricity	233
	11.8	Prestress moments due to cable curvature	234
	11.9	Prestress analysis by flexibility coefficients	236
	11.10	Prestress applied directly to space frame	238
		References	242
12.	**Harmonic analysis and folded plate theory**	244
	12.1	Introduction	244

12.2	Harmonic components of load, moment, etc	245
12.3	Characteristics of low and high harmonics	249
12.4	Harmonic analysis of plane decks	253
12.5	Folded plate analysis	255
12.6	Continuous and skew decks	260
12.7	Errors of harmonics near discontinuities	261
References		262

13. Finite element method — 263

13.1	Introduction	263
13.2	Two-dimensional plane stress elements	264
13.3	Plate bending elements	271
13.4	Three-dimensional plate structures and shell elements	275
13.5	Finite strips	276
13.6	Three-dimensional elements	278
13.7	Conclusion	279
References		280

14. Stiffnesses of supports and foundations — 281

14.1	Introduction	281
14.2	Substructures and bearings	282
14.3	Foundation stiffnesses	283
14.4	Stiffness moduli of soils	290
14.5	Stiffnesses from lateral earth pressures	291
14.6	Embankment movements	293
14.7	Integral bridges	294
References		299

Appendix A — 301
Product integrals. Functions of load on a single span.
Harmonic components

Appendix B — 305
Calculation of torsion constant for solid beams

Author Index — 307
Subject Index — 309

Acknowledgements

Many people helped me with this book. I am grateful to Gifford & Partners with whom I worked during the design of several of the bridges that form examples. In particular I gained much from interacting with: Malcolm Woolley, Maurice Porter, Ernest Pennells and Edmund Hollinghurst. My interest in bridge design first developed while I worked for Ove Arup and Partners under Robert Benaim. Later, John Blanchard and Peter Dunican encouraged me to write a book; Edgar Lightfoot gave me crucial advice to make it readable. Many people made comments on the first edition which led to improvements in the second edition. I also thank the organizations and individuals who provided photographs and are identified in the captions; my own photographs benefited from the guidance of Ernest Janes.

Finally, I would like to record my continuing gratitude to Michael Chrimes and the Staff of the Library of the Institution of Civil Engineers in London who provide a remarkably friendly and conscientious service to everyone who turns to them for assistance.

E.C.H.

Preface

This book describes the load distribution behaviour of steel and concrete bridge decks. The principles can also be applied to several other materials and deck-type structures. The book has been written to be intelligible to junior engineers who are interested in the physical characteristics of the different types of construction and who require detailed descriptions of some calculation methods. The book has also been written with consideration for the senior engineers leading design teams, to give them information about the range of analytical methods available and on some of their shortcomings. It has been assumed that the design and assessment of bridges are entrusted to experienced professional civil engineers, and that calculations are carried out under the direction of appropriately experienced and qualified supervisors. Users of this book are expected to draw upon other works on the subject including national and international codes of practice, and are expected to verify the appropriateness and content of information they draw from this book.

In this second edition, Chapter 1 has been enlarged to explain the intrinsic safety of some calculation methods when used in a systematic manner. Chapter 2 on beam decks and frames now includes demonstrations of the influence of foundation stiffnesses on the frame behaviour of a portal bridge and an arching structure. Chapter 3 on slabs now includes the analysis of shear-key decks. Chapter 4 on beam-and-slab bridges has been enlarged with new sections on 'torsionless design' which can simplify the design of some bridges, and on bracing of steel beams. New examples illustrate the grillage analysis of a composite bridge using AASHTO-type prestressed concrete girders, and of a composite steel deck. Chapter 5 on cellular structures includes additional comparisons between grillage and finite strip analyses.

A new Chapter 6 explains the distortion of box-girder bridges. Examples illustrate the analysis of box girders by beams-on-elastic-foundations, two types of grillage and space frame models. In recent years the author has made increasing use of simple space frame idealizations of three-dimensional structures, and Chapter 7 has been enlarged to demonstrate their wide range of applications. The discussion of shear lag and edge stiffening in Chapter 8 has additional comments on load redistribution. Chapter 9 on skew and curved decks illustrates the interaction of bending and torsion in a space frame analysis of a curved multispan bridge. Chapter 11 on temperature and prestress has been enlarged to demonstrate the use of space frame models for calculation of the effects of temperature and prestress. A new example illustrates the effects of post-tensioning on the torsion and bending of a curved multispan bridge. Prestress and other internal loads can be applied directly to the computer model, in the same manner as to the real structure, so that it is not necessary to calculate equivalent loads or to separate the effects into 'primary' and 'secondary'.

A new Chapter 14 has been included on the stiffnesses of supports and foundations. The stiffnesses of supports and foundations are becoming increasingly important as bridge engineers turn to integral bridge designs with no movement joints in order to reduce maintenance problems. A worked example illustrates the global analysis of an integral bridge, including deck, supports and ground. Appendix B explains the calculation of torsion stiffness.

The overall objectives of the second edition, like the first, have been to explain and demonstrate the underlying principles of different bridge types. Each designer can apply them to suit the particular bridges, computer programs and design codes relevant to his or her part of the world.

The book shows how complex structures can be analysed with physical reasoning and relatively simple computer models, and without complicated mathematics. In recent years the computer methods of grillage and space frame have become very popular and accessible as microcomputers and software have developed rapidly. The visual displays of modern programs can provide an engineer with a comprehensive picture and understanding of the behaviour of his structure. At the design stage this helps him to manipulate his design and so economize in the use of construction materials. During the assessments of old bridges he can examine alternative load paths with ease and so determine the reserves of strength as the structure changes. The improvements in facilities since the publication of the first edition now enable the author to analyse in one hour a deck which previously took several days.

Engineers, in general, have confidence in their calculations only when they can back them up with physical reasoning. For this reason this book concentrates on the physical reasoning that is necessary to translate prototype behaviour and properties into computer models, and vice versa. Most attention is paid to the simpler methods of grillage and space frame because they are more commonly used. With experience engineers are able to use physical reasoning and simple models for the design of relatively complex structures. However, since such experience involves comparisons of results of these simple methods with test results and solutions of more rigorous analyses, the principles of space frame, folded plate and finite element methods are described in later chapters. The only mathematics that is necessary for the majority of bridge deck designs is summarized in Chapter 2 and concerns simple beam theory that is covered in most university first year courses on civil and structural engineering.

Hand methods of analysis are also very useful and will remain essential for preliminary design, checks, and when the computer is not available. Initially the author greatly preferred such methods to the general use of the computer, and an early draft of this book concentrated on the subject. However with increased experience and responsibility, a complete change of attitude became necessary because the computer methods had the following advantages.

1. They are comprehensible to the majority of engineers, many of whom, though thoroughly competent, do not have the mathematical expertise in techniques, such as harmonic analysis, that are needed for accurate application of many hand methods to complicated structures.
2. They are applicable to the majority of bridge shapes with skew, curved or continuous decks and with varying stiffness from region to region. In contrast, hand methods are simple to use only for the few bridges which are rectangular in plan and simply supported.
3. They are also applicable, with shear flexibility, to a much wider variety of deck cross-sections.
4. They are checkable; it is much easier to check computer data and output distributions of forces than pages of hand calculations.
5. Finally, they are economical. With the development of very convenient and clear grillage programs computer data can be prepared, numerous load cases analysed, and the results processed in a much shorter time than the equivalent hand calculations can be carried out.

None the less, because hand methods are still very useful, some published techniques are reviewed in Chapter 10 and applications of rapid design charts are demonstrated.

The accuracy of any method of analysis for a particular structure is difficult to predict or even check. It depends on the ability of the model to represent three very complex characters: the behaviour of the material, the geometry of the structure, and the actual loading.

Construction materials, even when homogeneous, have properties differing widely from the elastic, or plastic idealizations. When incorporated in a structure they have innumerable variations of stiffness and strength owing to composition, and site and life histories. The analysis almost invariably simplifies the geometry of the structure of thick members to an assemblage of thin plates or beams. Numerous holes, construction joints, site imperfections and other details are ignored. Finally the design loadings for live load, temperature, creep, settlement and so on are idealizations based on statistical studies. It is unlikely that the critical design load will ever act on the structure even though it might be exceeded. For these reasons, large errors are likely whatever method of analysis is used. It is suggested that greater emphasis should be given to considering the physical behaviour of the structure and anticipating consequences of calculations being in error by more than 20% than to refining calculations in pursuit of the last 1% of apparent accuracy.

Dr E.C. 'Tim' Hambly
MACantab, PhD, FEng, FICE, FIStructE, MASCE.
Home Farm House, Little Gaddesden, Berkhamsted, Herts HP4 1PN.
September 1990

Notation

Superscripts
- ¯ average value or global variable
- ^ maximum value
- ′ relates to top slab of cellular deck
- ″ relates to bottom slab of cellular deck

Subscripts
- b   BEF-equivalent beam
- c   about centroid, or complete section
- e   of equivalent grillage member, or of effective flange
- f   of flange
- l   longitudinal
- M   due to bending
- S   due to shear
- T   due to torsion
- t   transverse
- u   undrained
- W   related to loading
- w   of web
- x, y, z axis of member, or moment, force or section property related to vertical bending of that member
- xx, yy, zz local axis for direction of force and associated shear area or about which moment acts
- I, II   principal values
- $1, 2, ..., n$ number of end, or slab edge, or support, or node, or beam, or harmonic

A area of, cross-section, or part section, or enclosed area
A_S equivalent shear area
a stiffness coefficient, or dimension, or harmonic coefficient
a_S equivalent shear area per unit width

BEF	beam-on-elastic-foundations
b	breadth, or stiffness coefficient, or harmonic coefficient
C	torsion constant
c	torsion constant per unit width, or stiffness coefficient
c	cellular stiffness ratio
D	flexural rigidity
d	depth, or thickness
E	Young's modulus
e	eccentricity of prestress
F	node force
f	flexural stiffness ratio
G	shear modulus
g	stiffness coefficient
H	abutment height
h	distance between midplanes of slabs in cellular decks
h', h''	distances of midplanes of top and bottom slabs from their common centroid
I	moment of inertia = second moment of area
i	moment of inertia per unit width
J	influence value
j	shear flexibility parameter
K, k	stiffness matrices
k	spring stiffness, or stiffness coefficient
L	span, or distance between points of contraflexure
l	length, or web, or 'beam' spacing, or BEF panel length
M	bending moment
M_{FE}	fixed end moment
m	moment per unit width, or modular ratio, or moment system in flexibility analysis
N	applied torque
n	harmonic number, or stiffness coefficient
O	origin
P	force, or prestress compression force, or passive soil force
q	BEF bracing stiffness
R	radius of curvature, or reaction
r	force matrices
r	shear flow
r	rotational stiffness ratio
S	shear force
S_{FE}	fixed end shear force
S_U	soil undrained strength
s	shear force per unit width, or distance around curved arc or midplane

T	torque
t	torque per unit width, or thickness
t	axis rotation transformation matrix
U	applied load in Ox direction
u	displacement in Ox direction (warping)
U, u	force or displacement matrices
V	applied load in Oy direction
v	displacement in Oy direction, or BEF out-of-plane shear/torsion
W	applied load in Oz direction (vertical downwards), or BEF deflection
w	displacement in Oz direction (vertical downwards), or BEF deflection
Ox	horizontal axis along span (except where given local direction)
X, x	release action in flexibility analysis
X	load matrix
Oy	horizontal axis transverse to span (except where given local direction)
y	horizontal distance of point to side of origin or neutral axis
Z	amplitude of harmonic component of vertical load
Oz	vertical axis downwards (except where given local direction)
z	vertical distance of point below origin or neutral axis
α	angle, or coefficient of thermal expansion, or $(n\pi/L)$
$\alpha_1, \alpha_2, \ldots$	coefficients of displacement field
β	BEF parameter
γ	shear strain, or soil density
Δ	increment
δ	flexibility coefficient, or displacement
ε	linear strain
θ	rotation, usually slope $\partial w/\partial x$, or inclination of prestress
ν	Poisson's ratio
σ	tension/compression stress
τ	shear stress
ϕ	rotation, usually $\partial w/\partial y$, or soil angle of friction, or creep factor
$\dot{\phi}$	twist = rate of change of ϕ with length

Units

The examples in this book use SI metric units.

Dimensions are generally expressed in metres (m) where
$$1\,\text{m} = 3.28\,\text{ft}$$
Forces are generally expressed in meganewtons (MN) where
$$1\,\text{MN} = 100\,\text{tons}.$$
Stresses and pressures are expressed in megapascals (MPa) or meganewtons per square metre (MN/m^2), which is the same as newtons per square millimetre (N/mm^2).
$$1\,\text{MPa} = 1\,\text{MN/m}^2 = 1\,\text{N/mm}^2 = 145\,\text{psi}.$$

1 Structural forms and calculation methods

1.1 INTRODUCTION

Bridge decks are developing today as fast as they have at any time since the beginning of the Industrial Revolution [1–4]. The diversity of sites is increasingly challenging the ingenuity of engineers to produce new structural forms and appropriate materials. Methods of analysis have developed equally rapidly, particularly with the use of computer methods. The accessibility of microcomputers is making it progressively easier for engineers to analyse bridges with complex cross-sections and complicated skew, curved and continuous spans. In the past a considerable amount of theoretical and experimental research was required to develop the design methods. Today, however, several have been developed to such usable form that, with an understanding of physical behaviour, designers can analyse complex decks without recourse to complicated mathematical theory.

This book concentrates on the simpler computer methods of calculation of grillage and space frame. Section 1.3 explains the merits and intrinsic safety of the methods when used in a systematic manner. One of the most useful aspects of the methods is that each part of a structure is represented by equivalent beam elements. Most structural engineers have an intuitive feel for how beams react to the various forces of bending, shear and torsion, and they can use physical reasoning. A simple physical model, such as illustrated in Fig. 1.2, is often more reliable than complicated calculations as a guide to the size and directions of forces and displacements.

Section 1.2 reviews and categorizes the principal types of bridge deck that are currently being used, and refers to the analytical techniques demonstrated in later chapters. The types of bridge deck are divided into beam, grid, slab, beam-and-slab and cellular, to differentiate their individual geometric and behavioural characteristics. Inevitably many decks fall into more than one category, but they can usually be analysed

2 Structural forms and calculation methods

Fig. 1.1 Wrought iron box girders of 140 m span of Britannia Bridge, Wales, 1849; designed by Robert Stephenson. Lithograph from Cyclopaedia of Useful Arts by C. Tomlinson, 2nd edn, 1866. Photograph courtesy of Elton Engineering Books.

by using a judicious combination of the methods applicable to the different types.

1.2 STRUCTURAL FORMS

1.2.1 Beam decks

A bridge deck can be considered to behave as a beam when its length exceeds its width by such an amount that when loads cause it to bend and twist along its length, its cross-sections displace bodily and do not change shape, as shown in Fig. 1.3.

The most common beam decks are footbridges, either of steel, reinforced concrete or prestressed concrete. They are often continuous over two or more spans. Many long-span bridges behave as beams because the dominant load is concentric so that the distortion of the cross-section under eccentric loads has relatively little influence on the principal bending stresses.

The analysis of bending moments and torsions in continuous beam decks is discussed in Chapter 2. Frame action is also discussed for decks

Structural forms

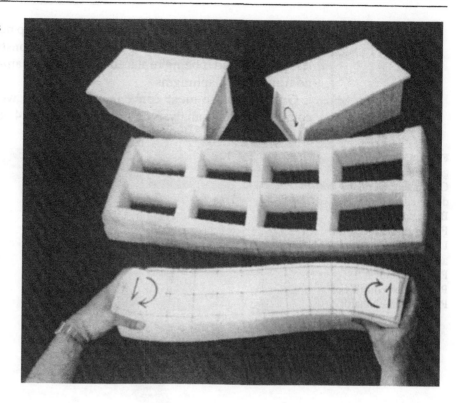

Fig. 1.2 Physical models for understanding distributions of forces and displacements.

in which the stiffness and geometry of the supports have a significant influence on behaviour.

1.2.2 Grid decks
The primary structural member of a grid deck is a grid of two or more longitudinal beams with transverse beams (or diaphragms) supporting the running slab. Loads are distributed between the main longitudinal beams by the bending and twisting of the transverse beams, as shown in

Fig. 1.3 Beam deck bending and twisting without change of cross-section shape.

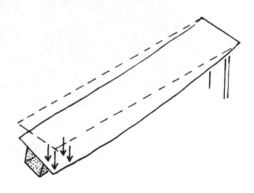

Fig. 1.5. Because of the amount of workmanship needed to fabricate or shutter the transverse beams, this method of construction is becoming less popular and is being replaced by slab and beam-and-slab decks with no transverse diaphragms.

Grid decks are most conveniently analysed with the conventional computer grillage analysis described in Chapter 4. The analysis in effect

Fig. 1.4 Concrete box girder Kocher Viaduct, Geislingen, Germany; designed by Prof. Peter Bonatz of Wayss & Frytag, Frankfurt, with Leonhardt and Andra as advisor and Prufingenieur. Photograph courtesy of Prof. Fritz Leonhardt.

Fig. 1.5 Load distribution in grid deck by bending and torsion of beam members.

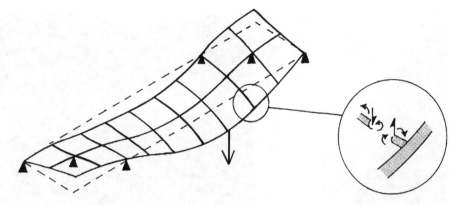

sets out a set of simultaneous slope–deflection equations for the moments and torsions in the beams at each joint and then solves the equations for the load cases required. Grid decks have also been analysed for many years by hand methods which are summarized in Chapter 10. However, such methods are decreasing in popularity as computer methods become simpler and more versatile.

1.2.3 Slab decks

A slab deck behaves like a flat plate which is structurally continuous for the transfer of moments and torsions in all directions within the plane of the plate. When a load is placed on part of a slab, the slab deflects locally in a 'dish' causing a two-dimensional system of moments and torsions which transfer and share the load to neighbouring parts of the deck which are less severely loaded, as shown in Fig. 1.6. A slab is 'isotropic'

Fig. 1.6 Load distribution in slab deck by bending and torsion in two directions.

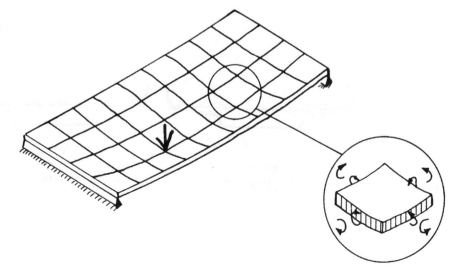

6 Structural forms and calculation methods

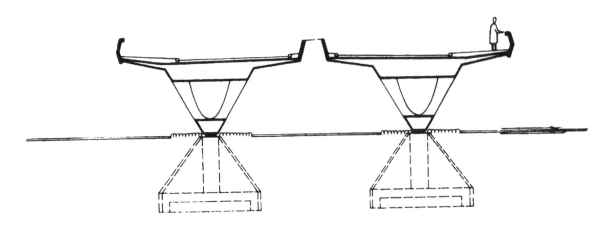

Fig. 1.7 Concrete slab decks of Western Bank Bridge, Sheffield, England; designed by Ove Arup & Partners. Photograph courtesy Ove Arup & Partners.

when its stiffnesses are the same in all directions in the plane of the slab. It is 'orthotropic' when the stiffnesses are different in two directions at right angles.

Concrete slab decks are commonly used where the spans are less than 15 m (50 ft). If the deck is cast *in situ* it is general practice to consider it as isotropic, even though the reinforcement may not be the same in all directions.

When it is inconvenient to support the deck on falsework during construction, the slab is often built compositely with reinforced concrete cast *in situ* between previously erected beams, as shown in Fig. 1.8(b). The beams of precast concrete or steel have a greater stiffness longitudinally than the *in situ* concrete has transversely; thus the deck is orthotropic.

For spans greater than 15 m (50 ft), the material content and dead load of a solid slab become excessive and it is customary to lighten the structure by incorporating voids of cylindrical or rectangular cross-section near the neutral axis. These are shown in Fig. 1.8(c) and (d). If the depth and width of the voids are less than 60% of the overall structural depth, their effect on the stiffness is small and the deck behaves effectively as a plate. Voided slab decks are frequently constructed of concrete cast *in situ* with permanent void formers, or of precast prestressed concrete box beams post-tensioned transversely to ensure transverse continuity. If the void size exceeds 60% of the depth, the deck is generally considered to be of cellular construction with a different behaviour, as is described later.

Rigorous analysis of most slab decks is not possible at present. However, slab decks can be conveniently analysed using the computer grillage as described in Chapter 3. In this method the continuous slab is represented by an equivalent grid of beams whose longitudinal and transverse stiffnesses are approximately the same as the local plate stiffnesses of the slab. This analogy is not the closest available, but it has been found to agree well with more rigorous solutions and is generally

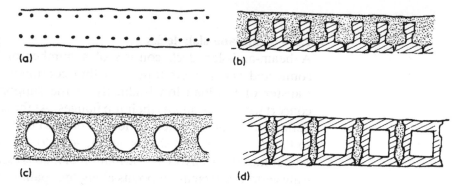

Fig. 1.8 Slab decks: (a) solid; (b) composite of *in situ* concrete in-filling precast beams; (c) voided; and (d) voided of precast box beams post-tensioned transversely.

Fig. 1.9 Differential deflections of beams in shear-key deck resisted by torsion of beams.

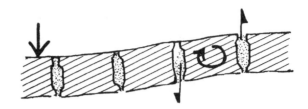

accepted as sufficiently accurate for most designs. One of the more exact methods, which is mentioned in Chapter 13, is performed using a computer finite element program. In this method the deck is notionally split up into a number of elements, frequently triangular, for each of which approximate plate bending equations can be derived and solved (as opposed to the beam equations used in the grillage). The method is much more complicated and expensive than the grillage, and since for slab decks it does not generally produce significantly different results, the grillage is normally used in preference. The hand methods of Chapter 10 also provide a convenient method of analysis for the design of slab decks with simple plan geometry. However, as mentioned for grid decks, computer methods are becoming easier to use and provide more information.

One type of deck which does not fit neatly into any of the main categories is the 'shear-key' deck. A shear-key deck is constructed of contiguous prestressed/reinforced concrete beams of rectangular or box sections, connected along their length by *in situ* concrete joints. It is not prestressed transversely and thus is not fully continuous for transverse moments. The main application is for bridges constructed over busy roads or railways, where it is convenient to erect the beams overnight and then complete the jointing without disturbance to the traffic below. Although such decks have little or no transverse bending stiffness, distribution of loads between beams still takes place because differential deflection of the beams is resisted by the torsional stiffness of the beams and a vertical shear force is transferred across the keyed joints. The analysis of such decks is described in Chapter 3.

1.2.4 Beam-and-slab decks

A beam-and-slab deck consists of a number of longitudinal beams connected across their tops by a thin continuous structural slab. In transfer of the load longitudinally to the supports, the slab acts in concert with the beams as their top flanges. At the same time the greater deflection of the most heavily loaded beams bends the slab transversely so that it transfers and shares out the load to the neighbouring beams. Sometimes this transverse distribution of load is assisted by a number of transverse diaphragms at points along the span, so that deck behaviour is more similar to that of a grid deck. However, the use of diaphragms is becoming less popular because of the construction problems they cause and because their localized stiffnesses attract forces which can cause unnecessary stress concentrations. Beam-and-slab construction has the advantage over slab that it is very much lighter while retaining the necessary longitudinal stiffness. Consequently it is suitable for a much wider range of spans, and it lends itself to precast and prefabricated construction. Occasionally, the transverse flexibility can be advantageous; for example, it can help a deck on skew supports to deflect and

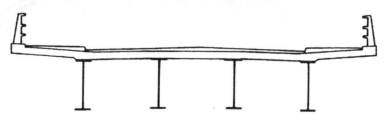

Fig. 1.10 Steel composite beam-and-slab bridge at Newburgh, Scotland; designed by Grampian Regional Council. Photograph courtesy Grampian Regional Council.

Fig. 1.11 (a) Contiguous beam-and-slab deck and (b) slab of contiguous beam-and-slab deck deflecting in smooth wave.

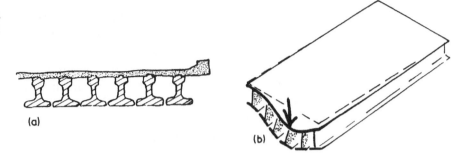

twist 'comfortably' under load without excessively loading the nearest supports to the load or lifting off those further away.

Beam-and-slab decks can be divided into two main groups: those with the beams at close centres or touching are referred to as 'contiguous beam-and-slab', while those with beams at wide centres are referred to as 'spaced beam-and-slab'. The most common form of contiguous beam-and-slab, shown in Fig. 1.11, comprises precast prestressed concrete inverted T-beams supporting a cast *in situ* reinforced concrete slab of about 200 mm (8 in) thickness. When a load is placed on part of such a deck, the slab deflects in a smooth wave so that for load distribution its behaviour can be considered similar to that of an orthotropic slab with longitudinal stiffening.

Spaced beam-and-slab decks shown in Fig. 1.12 commonly have the beams at about 2 m (6 ft) to 3.5 m (12 ft) centres. Decks have been designed with precast prestressed concrete beams or steel beams supporting a concrete slab. Numerous variations of construction have

Fig. 1.12 Spaced beam-and-slab decks with steel I-beams and (a) concrete slab (b) steel 'battledeck' running slab.

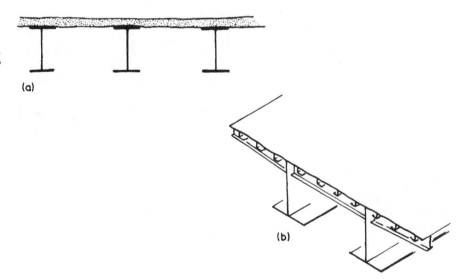

been employed for the concrete slab ranging from totally cast *in situ* to very large precast panels connected to the beams by the minimum of *in situ* concrete in the joints. Sometimes the running slab on steel beams consists of a steel 'battledeck' which is constructed of a stiffened steel plate of as little as 12 mm (½ in) thickness. When a load is placed above one beam of a spaced beam-and-slab deck, the slab does not necessarily deflect transversely in a single wave but sometimes in a series of waves between beams. This is particularly the case if the beams have high torsional stiffnesses, as do box beams, when the beams may twist little so that the slab deflects in a series of transverse steps as shown in Fig. 1.13. Such behaviour is different from that of orthotropic slabs and it is advisable for the analytical model to have its high longitudinal beam stiffnesses correctly positioned across the deck. If the various stiffnesses cannot be correctly located, the load distribution analysis is sometimes backed up with a plane frame analysis of the cross-section to study its transverse bending and distortion.

The extreme form of spaced beam-and-slab decks can have as few as two spine beams at more than 12 m (40 ft) centres. Solid concrete spines at more than 7 m (24 ft) are rare, but twin-spine concrete and steel box-girder decks are not uncommon. While the bending and torsional behaviour of the spines must be considered as described below for cellular decks, the distribution of loads between spines is essentially beam-and-slab.

Beam-and-slab decks are most conveniently analysed with the aid of conventional computer grillage programs, as mentioned for slab decks above. The application of this method to these decks is described in Chapter 4. This method is generally accepted as sufficiently accurate for design, but it ignores possible high horizontal shear forces in the slab resulting from differences in the shortening of the top fibres of adjacent beams subjected to different bending deflections. Methods for assessing these forces are described in Chapter 7.

Fig. 1.13 Spaced beam-and-slab deck deflecting in a series of steps or waves.

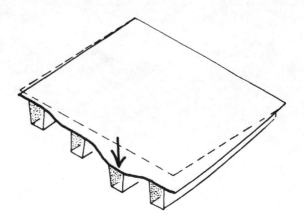

1.2.5 Cellular decks

The cross-section of a cellular or box deck is made up of a number of thin slabs and thin or thick webs which totally enclose a number of cells. These complicated structural forms are increasingly used in preference to beam-and-slab decks for spans in excess of 30 m (100 ft) because in addition to the low material content, low weight and high longitudinal bending stiffness they have high torsional stiffnesses which give them better stability and load distribution characteristics. The use of box decks has been particularly spectacular in recent years for long, high spans, where falsework is inappropriate, and the deck has been erected in elements as a beam cantilevering out from supports or the deck has been constructed and launched across the piers from an abutment. Cantilever construction is less popular with beam-and-slab decks because large trusses are usually needed temporarily to provide

Fig. 1.14 Steel box girder composite bridge at Nashville, Tennessee, USA; designed by State of Tennessee Department of Transportation. The 54 m centre span has short end spans built into the abutments to form an integral bridge with no movement joints between embankments. Photograph courtesy of George Hornel, State of Tennessee.

Fig. 1.15 Concrete box girder deck of segmental construction at East Moors Viaduct, Cardiff, Wales; designed by Robert Benaim & Associates. Photographs courtesy of Shepherd, Hill Ltd and Robert Benaim & Associates.

Fig. 1.16 Curved steel box girders of Yoshima Loop Bridge, in front of Kitabisan and Minamibisan Seto suspension bridges, Japan: designed by Honshu-Shikoku Bridge Authority. Photograph courtesy of Yokogawa Construction Company.

torsional stiffness to the incomplete deck. To describe the behaviour of cellular decks it is convenient to divide them into multicellular slabs and box-girders.

Multicellular slabs are wide shallow decks with numerous large cells. The cross-sectional shape does not lend itself to precast segmental construction, and construction is usually *in situ* concrete or contiguous precast box beams or top hat beams with large voids. For spans up to 36 m (120 ft), the cells of *in situ* concrete decks frequently consist of large cylindrical voids of diameter exceeding 60% of the structural depth. For longer spans such a deck would be too heavy, and the cross-section is lightened by making cells rectangular and enlarging them so that they fill most of the cross-section with the top and bottom slabs as thin as 200 mm (8 in) and 150 mm (6 in), respectively. When a load is placed on one part of such a deck, the high torsional stiffness and transverse bending stiffness of the deck transfer and share out the load over a wide area. The distribution is not as effective as that of a slab since the thin top and bottom slabs flex independently when transferring vertical shear forces between webs, and the cross-section is said to 'distort' like a Vierendeel truss in elevation, as shown in Fig. 1.17. Such distortion can be reduced by incorporating transverse diaphragms at various points along the deck but, as with beam-and-slab decks, their use is becoming less popular except at supports where it is necessary to transfer the vertical shear forces between webs and bearings.

Box-girder decks have a cross-section composed of one or a few large cells, the edge cells often having triangular cross-section with inclined outside web. Frequently the top slab is much wider than the box, with the edges cantilevering out transversely. Excessive twisting of the deck under eccentric loads on the cantilevers is resisted by the high torsional stiffness of the structure. Small- and medium-span concrete box-girders are usually cast *in situ* or precast in segments which are erected on falsework or launching frame and prestressed. Large spans are often of segmental cantilever construction with the segments precast or cast *in situ* on a travelling form prior to being stressed back on to the existing structure. Steel box-girders are also frequently constructed as segmented cantilevers, but sometimes they are prefabricated in long lengths, weighing several thousand tons, which are lifted into position by climbing jacks and connected together. Some box-girders have been

Fig. 1.17 Cell distortion in multicellular decks.

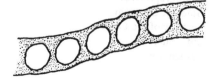

erected by constructing them at one abutment and launching them in stages over the piers.

The method of analysis most appropriate to a particular deck depends on the complexity of the structural form. If the deck has none or few transverse diaphragms a computer shear flexible grillage is adequate, as demonstrated in Chapter 5. In this method the deck is simulated by a grid of beams, as before, but the beams are given the high torsional stiffnesses of the cellular deck, and the slope–deflection equations take into account shear deformation in the beams. By attributing very low

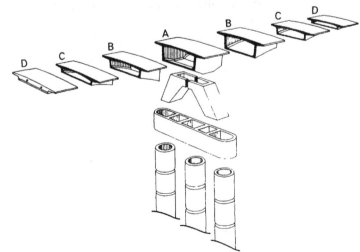

Fig. 1.18 Concrete box cantilever spans of Oosterschelde Bridge Holland (1965); designed and built by consortium van Hattum en Blankenvoort NV and W.V. Amsterdamsche Ballast Maatschappij. Photograph E.C. Hambly. Diagram courtesy of the Cement and Concrete Association.

shear stiffnesses to the transverse beams, their deformation in shear can be made to simulate cell distortion. However, the distortion of an unbraced single cell box-girder is much more complicated, as is explained in Chapter 6. The top and bottom flanges deflect sideways as the box distorts under eccentric loading. Grillage methods can be used, but the analysis may not give a sufficiently detailed picture of the flexural and membrane stresses in the plate elements, and, for this, additional three-dimensional analyses may be necessary. Space frame analyses, described in Chapter 7, have proved reliable and are liked by several design engineers because the deck is represented by an easily understood physical structure. If the deck has uniform cross-section from end to end and few transverse diaphragms, folded plate analysis (described in Chapter 12) probably provides the most accurate method. On the other hand, if the deck has complicated variations in section and numerous diaphragms, a finite element analysis (described in Chapter 13) may be necessary. This method can be complicated and expensive and it is often found most convenient to use it just to study stress flows in small parts of a structure while simpler methods are used to investigate the load distribution behaviour of the deck as a whole.

1.3 SAFETY OF METHODS

1.3.1 Equilibrium for safe design

An essential objective of any load distribution analysis, for the design of a structure, is to derive a system of forces which satisfies **equilibrium** at every point between internal forces and applied loads. The importance of this equilibrium of calculated forces cannot be overstated for the reasons explained below.

The calculation methods in this book are all based on elastic theory in which it is assumed that the deformations of the structures, and components, are linearly proportional to the applied loads. At the design stage of a bridge it is seldom practicable to predict precisely the collapse behaviour, because the materials have either a non-linear behaviour near failure, or have uncertain strengths. None the less it is possible to use elastic theory to produce a design which is safe and economic.

The shortcomings of elastic theory were overcome to a certain extent by plastic theory which was developed for the design of structures made with yielding materials, such as steel and concrete. Unfortunately, plastic methods are not as easy to use as elastic methods for complicated structures, and simple computer programs based on plastic theory are not readily available for bridge design. However plastic theory includes a safe design theorem, or lower bound theorem [5], that provides a very useful theoretical safety net for design by elastic methods.

The safe design theorem, based on plastic theory, can be stated as follows.

> *A structure will be safe when subjected to a design loading if the forces and stresses calculated throughout the structure are in equilibrium with each other and with the design loading, and if they do not anywhere exceed the yield strengths of the materials.*

The essential requirements of the theorem are:

1. the calculated forces and stresses should be everywhere in equilibrium;
2. the materials should be ductile; i.e. able to yield without loss of strength prior to failure;
3. the structure should not lose strength by buckling.

It is not necessary that the calculated system of forces should be exact predictions of the system that will exist in the real structure. It is only necessary that the calculated system of forces satisfies equilibrium at every point. In practice, it is impossible to predict the exact system of forces that will exist, and consequently the safe design theorem provides a very useful reassurance. It is inevitable that the forces calculated at the design stage will in some places be overestimates of the forces in the real structure and elsewhere underestimates. Ductility of materials ensures that if a component is actually subjected to forces exceeding the calculated values, the material can yield without losing strength, and so redistribute load to regions that are loaded less than calculated. However, the safe design theorem does not apply if buckling can intervene so that strength is lost during load redistribution.

The safe design theorem implies that a design should be safe even when minor mistakes have been made in the estimation of component stiffnesses, as long as equilibrium is satisfied and every component has ductility and can carry the force calculated for it. However, if the stiffness of a component is overestimated with the result that the calculated force in the component is incorrectly high, the component must be designed for the high force or else the analysis must be repeated. It would be incorrect to recalculate a reduced force or stress in isolation, unless compensating increases are accounted for elsewhere.

1.3.2 Systematic use of global and local analyses

The analysis of a complex structure is greatly simplified if the calculations are subdivided into levels of global, local and detailed analyses, thus:

1. the global analysis examines the distribution of forces and deflections along the spans of a multispan bridge, and across the width of a wide bridge;
2. the local analysis examines the distribution of forces and deflections around one beam or part of a span;
3. the detailed analysis examines the stresses at particular points in a part of the structure.

This subdivision of the analytical process takes advantage of St Venant's principle [6], which states that

if the forces on a small portion of an elastic body are replaced by another statically equivalent system of forces on the same portion of the body, this redistribution of loading produces substantial changes in the stresses locally but has negligible effect on the stresses at distances which are large in comparison with the dimensions of the portion on which the forces are changed.

In effect each of the levels of analysis can be undertaken independently of the geometric details of more localized analyses as long as the analyses are undertaken for statically equivalent systems of forces.

The statical equivalence of the forces in the different analyses is essential not only to comply with St Venant's principle but also to take advantage of the safe design theorem outlined in Section 1.3.1. The calculated system of forces and stresses in the different levels of analysis can be kept in equilibrium between analyses and within each analysis if the forces on the boundary of each analysis are statically equivalent to the forces on the equivalent part of the other analyses.

The boundary conditions of a local analysis, or a detailed analysis, might be derived from either the forces or the displacements output from the next larger level of analysis. Ideally, the stiffnesses of the analyses are similar so that the displacements of the different analyses are compatible at the same time that forces are in equilibrium. However, in general it is impracticable to make the stiffnesses exactly the same, so then the boundary conditions of the local analysis should be made with a system of forces which is statically equivalent to the forces in the same region of the global analysis rather than with compatible displacements.

1.3.3 Serviceability and fatigue
The safe design theorem relates to calculations for the ultimate strength of a structure. The elastic methods of this book are also applicable to calculations for serviceability and working conditions, and for fatigue under fluctuating loads. Many structures behave elastically under the

low-amplitude cyclic loading which can cause fatigue damage. However, fatigue strength calculations do not benefit from the safety net provided by the safe design theorem since no ductile redistribution occurs. The accurate prediction of local stresses depends on the accurate prediction of relative stiffnesses throughout the structure. Calculated stresses and fatigue lives will be realistic only if the spread of stiffnesses is realistic.

The serviceability and working conditions can be the most difficult to model since bridge structures, and the ground supporting them, experience long-term changes due to creep, shrinkage and settlement. The long-term behaviour under dead load can differ markedly from the short-term behaviour under live load. The uncertainties of long-term effects have often been avoided in the past by making structures statically determinate with numerous movement joints. However, widespread maintenance problems, associated with water and salts penetrating the joints, are leading to pressure on designers to avoid movement joints when possible. It is likely that more structures will be statically indeterminate in the future. When long-term effects cannot be predicted precisely, it is usually possible to produce calculations for upper and lower bound estimates. Computer programs are now so easy to use that a structure can be reanalysed quickly for a variety of stiffnesses, corresponding to different short-term and long-term conditions.

1.3.4 Simple methods for safe design

This book champions the simpler methods of computer analysis using grillage, or space frame, for the following reasons.

1. The grillage and space frame methods are able to calculate systems of forces which are in equilibrium with themselves and with applied loads throughout the structure. Well-written computer programs report on a check that equilibrium has been achieved after each calculation, and so benefit from the safety net of the safe design theorem.
2. Grillage and space frame methods calculate beam forces, as opposed to local stresses. Beam forces can be used both for strength design of structural components, in which local stresses may be non-linear, as well as for permissible stress design with linear distributions.
3. Calculated beam forces can be compared directly with strength formulae for beams in codes of practice. In contrast, codes of practice provide relatively little guidance on conditions of stress at a point.
4. The computer should calculate automatically the forces in every

component that is to be designed, so making it easy to identify the most highly loaded parts.
5. The simpler an analysis is, the easier it is to verify that there are no arithmetic errors in the data, and the easier it is to check that the results are sensible.

It should not be taken for granted that the forces calculated by a computer program, or by a hand method, are automatically in equilibrium with applied loads. Not only can the calculation have a fault, but some methods, even when faultless, do not involve a rigorous check of equilibrium. Finite element methods are capable of representing in great detail the three-dimensional geometry of a structure and the behaviour of materials. Unfortunately, the realism and detail of some types of element is achievable only with approximations in numerical methods. The numerical methods do not necessarily involve a rigorous check of equilibrium of output stresses, although they may approximate to it. Grillage and space frame methods are less realistic at modelling the geometry and material properties, but most programs automatically involve equilibrium of output forces. A compromise is always involved. A very detailed finite element analysis is not necessarily more reliable than a simpler grillage or space frame.

No method of calculation provides an exact prediction for the behaviour of a complex structure. The best estimates are usually obtained by the judicious use of more than one method of calculation. Confidence is gained by comparing the results of different methods, backed up by physical reasoning. Engineers have to use their own judgment and should not rely blindly on any specific formula or method.

1.4 HOOKE'S LAW AND YOUNG'S MODULUS

Elastic methods of calculation are based on Hooke's law that the deformation of a body varies linearly with the force on it. Robert Hooke first enunciated his law in 1676 as the cipher 'ceiiinossstttuu' which rearranged gives 'ut tensio sic uis' – 'as the extension so the force'.

In 1678 Hooke published his theory in more detail in his 'Lecture De Potentia Restitutiva, or of Spring, Explaining the Power of Springing Bodies' [7]. In lectures at Gresham College in London he explained his theories not just for springs, but also for building materials, structures and air. He also considered the energy stored by springs, the isochronous frequencies of vibration, and explained pressures in terms of vibratory momentum of particles. Fig. 1.19 reproduces Hooke's illustration of the flexing of a beam.

Robert Hooke was a brilliant structural engineer and architect as well as a scientist of genius. In 1666 after the Great Fire of London he was

Fig. 1.19 Robert Hooke's illustration of the bending of a beam in his paper of 1678 explaining the power of springing.

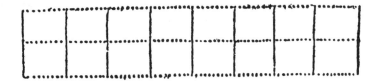

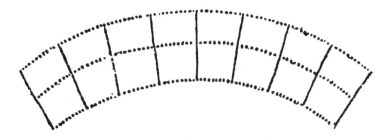

appointed one of the commissioners for the rebuilding and Surveyor of the City of London. He worked alongside Christopher Wren who was also a commissioner. While Wren was responsible for the King's works, Hooke was responsible for many of the City's, but they helped each other. One of Hooke's most remarkable structural achievements was devising the design method for the structural dome of St Paul's. This dome of brick is very thin and relies on its three-dimensional shape for its strength.

Robert Hooke realized that the stiffnesses of materials differed. However it was not until 1802 that the characteristic stiffness of each material was defined by Thomas Young; so that it is now known as Young's modulus. Thomas Young was also a scientist of extraordinary genius and he worked in many of the subjects that were studied by Hooke. Engineers since the time of Young have simplified his definition to mean the ratio of stress to strain. But he described it as follows in a lecture to the Royal Institution in London (published in 1807 [8])

> 'We may express the elasticity of any substance by the weight of a certain column of the same substance, which may be denominated by the modulus of its elasticity, and of which the weight is such, that an addition to it would increase it in the same proportion as the weight added would shorten, by its pressure, a portion of the substance of equal diameter.'

Thomas Young's activities included advising the government of the day on Telford's proposals for a single-arch bridge across the Thames, and he wrote the section in Encyclopaedia Britannica on bridges which amounted to more than 20 000 words.

REFERENCES
1. Leonhardt, F. (1982) *Bridges*, Architectural Press, London.
2. Beckett, D. (1969) *Great Buildings of the World: Bridges*, Hamlyn, London.
3. Lee, D.J. and Richmond, B. (1988) 'Bridges', in *Civil Engineer's Reference Book*, (ed. L.S. Blake) Butterworths, Guildford.
4. Liebenberg, A.C. (1983) 'Bridges', in *Handbook of Structural Concrete*, (eds F.K. Kong et al.) Pitman, London.
5. Horne, M.R. (1979) *Plastic Theory of Structures*, Pergamon, Oxford, 2nd edn.
6. St Venant, B. de (1855) Mem. des savants étrangers, Vol. 14.
7. Hooke, R. (1678) *Lectures De Potentia Restitutiva, or of Spring, Explaining the Power of Springing Bodies*, The Royal Society, London.
8. Young, T. (1807) *A Course of Lectures on Natural Philosophy and the Mechanical Arts*, The Royal Institution, London.

2 Beam decks and frames

2.1 INTRODUCTION
This chapter is concerned with bridge decks which can be thought of as beams or frames. After a review of the forms of deck there is a summary of the simple elastic theory of bending and torsion of beams. This summary is intended to be a brief reference to remind the reader of the basic theory on which much of the load distribution analysis of this book is based and which is used for a wide variety of different structures. The study is far from comprehensive, so several references are given to books covering the subject more fully.

2.2 TYPES OF BEAM DECK
Figure 2.1(a)–(d) shows typical forms of beam deck. The simplest (a) is simply supported on three bearings so that it is statically determinate for bending and torsion. With four bearings in two right pairs as in (b), the deck is statically determinate for bending but not for torsion. Multiple span bridges are often built with spans simply supported as in (c) or continuous as in (d). The statical determinacy in (c) is advantageous when the stability of supports is uncertain due to subsidence. However, in general, multiple span bridges benefit from having a continuous deck because the elimination of movement joints removes a major cause of maintenance problems from penetration of dirt, water and de-icing salts, which corrode substructures. Integral bridges which have no movement joints are discussed in Chapter 14. Continuous decks also benefit from being more slender structures than simple spans.

Numerous other articulation and span arrangements are possible as shown in Fig. 2.2. Often the articulation is changed during the construction stages so that it is common for a deck to be statically determinate as beam or cantilever during construction and then made partially or totally continuous for live loads and long-term movements.

The bearings of a beam deck can be placed on a skew. If they are also far apart and incompressible, moment and torsion interact significantly at the supports. Analysis of such interaction can be cumbersome by hand, and since it is often not possible with a continuous beam program it may be necessary to use a two-dimensional computer analysis as described in later chapters. In a curved beam deck, moments and torsions interact throughout its length and although hand calculations are possible, the two-dimensional analysis described in Chapter 9 is often more convenient.

It is frequently appropriate to use a continuous beam analysis during the design of a multiple span wide deck with deformable cross-section if the supports have little or no skew and the deck is near straight. Under these conditions the distribution of total moment and shear force on cross-sections along the span is virtually the same as that for a single beam with the same total stiffness carrying a load with the same longitudinal disposition. A continuous beam analysis can be used to analyse longitudinal distributions of total moments and shear forces due to dead load, construction sequence, prestress, temperature, live load, etc. Then a two-dimensional grillage can be used to determine the

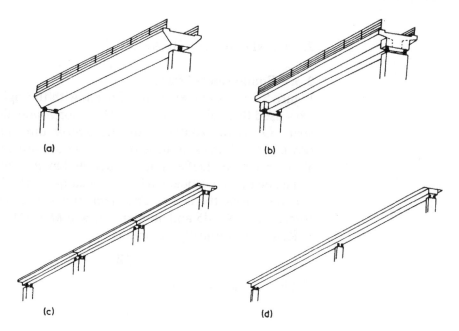

Fig. 2.1 Statical determinacy of bridges. (a) determinate for bending and torsion; (b) determinate for bending only; (c) multiple span simply supported determinate for bending; and (d) multiple span continuous indeterminate.

26 Beam decks and frames

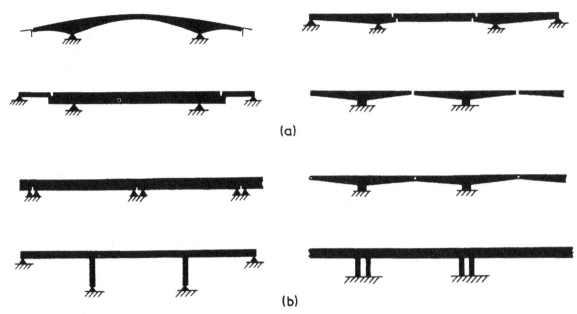

Fig. 2.2 Further examples of determinacy of bridges in bending. (a) determinate and (b) indeterminate.

transverse distribution of moments, shear forces and accompanying torsion due to lateral disposition of loads and stiffnesses.

2.3 BENDING OF BEAMS

2.3.1 Equilibrium of forces

Figure 2.3 shows an element of beam with the right-handed system of axes used throughout the book: Ox is horizontal along the direction of span, Oy is horizontal in the transverse direction, Oz is vertically downwards. Forces in these three directions are denoted by U, V, and W, respectively. Deflections are denoted by u, v, and w.

The element of a beam in Fig. 2.3 is subjected to a vertical load dW. W can vary along the beam. The element is held in equilibrium by shear forces S and $S + dS$ and moments M and $M + dM$ on the end faces.

Resolving vertically we obtain

$$dW = -dS. \tag{2.1}$$

Taking moments about Oy,

$$S dx = dM$$

or

$$S = \frac{dM}{dx}. \tag{2.2}$$

Fig. 2.3 Element of beam.

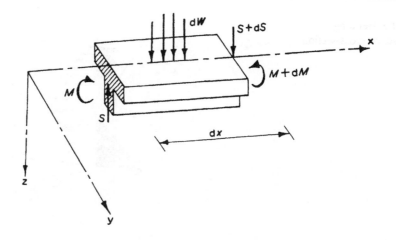

As shown in Fig. 2.3, positive shear forces here 'rotate' clockwise and positive moments cause sagging.

When a simply supported beam supports a load such as the point load in Fig. 2.4(a), shear force and bending moment at any point along the beam can be found from a consideration of equilibrium. From these, a shear force diagram and bending moment diagram can be drawn as shown in Fig. 2.4(b) and (c). In these diagrams, positive shear force and moment are plotted downwards. Similar diagrams for other common design loads are included in Appendix A.

2.3.2 Stress distributions

In the simple theory of elastic bending of beams it is assumed that the plane sections remain plane and that the beam is composed of discrete

Fig. 2.4 (a) Load; (b) shear force; and (c) moment diagram.

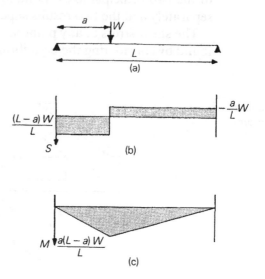

Fig. 2.5 Bending stresses on cross-section.

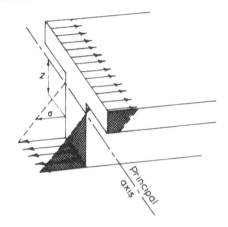

linear fibres in which the longitudinal bending stress σ is proportional to the longitudinal strain ε in that fibre. From these assumptions it can be shown [1–4] that, at any point on a cross-section, σ is proportional to the distance z of the point from the neutral axis, which passes through the centroid of the section. This is illustrated in Fig. 2.5. For flexure about a principal axis of the section

$$\frac{\sigma}{z} = \frac{M}{I} = \frac{E}{R} \qquad (2.3)$$

where M is the total moment on section, I is the second moment of area or 'moment of inertia' of section about the neutral axis, E is Young's modulus, and R is the radius of curvature of the beam due to flexure.

If a beam is subjected to loads which are not normal to a principal axis of the section then the loads must be resolved into components normal to the two principal axes. Bending about each axis is then considered separately and the two results superposed.

The shear stress at any point on a cross-section such as a in Fig. 2.6 is found by considering the equilibrium of that part $abcd$ of an element of

Fig. 2.6 Forces on slice along element of beam.

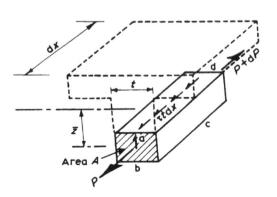

beam cut off by slice *ad* along the beam through *a*. The longitudinal tension force *P* on end *ab* is expressed as

$$P = (\text{average tension stress } \bar{\sigma} \text{ on } ab) \times (\text{area } A \text{ of the face } ab).$$

From equation (2.3),

$$\bar{\sigma} = \frac{M\bar{z}}{I}$$

where $\bar{z}$ is the distance of the centroid of area A from the centroid of the whole section.

$$\therefore \quad P = \frac{MA\bar{z}}{I}.$$

Between ends *ab* and *cd* of the element, *P* changes by d*P*:

$$dP = \frac{dMA\bar{z}}{I}.$$

Resolving longitudinally for load on *abcd*,

$$\tau t\,dx = dP = \frac{dMA\bar{z}}{I}$$

where τ is the longitudinal shear stress along cut *ad*. Replacing d*M*/d*x* by *S* we obtain

$$\tau t = \frac{SA\bar{z}}{I}. \tag{2.4}$$

Equilibrium of shear stresses at any point requires that they should be complementary in orthogonal directions. Hence equation (2.4) applies to shear stresses at *a* in both directions *ab* and *ad*.

The distribution of bending shear stress within a box deck, such as in Fig. 2.7, can be found with equation (2.4), but the beam must first be notionally cut longitudinally along its vertical axis of symmetry (on which there are no longitudinal shear stresses). Each side of the deck is then assumed to carry half of the bending shear force with second

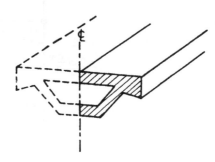

Fig. 2.7 Halving of box deck for analysis of shear stress.

moment of area I equal to half the total second moment of area of the deck cross-section. If the section is not symmetric, the top and bottom flanges should be cut in the manner described in Section 5.4.1 so that the different sections associated with each web have their individual neutral axes at the same level as the neutral axis of the whole section.

2.3.3 Slopes and deflections
For small deflections

$$R = -\frac{1}{\dfrac{d^2 w}{dx^2}}.$$

Hence we can write the second half of equation (2.3) in the form

$$\frac{d^2 w}{dx^2} = -\frac{M}{EI} \tag{2.5}$$

which can be integrated to give

$$\frac{dw}{dx} = -\int \frac{M}{EI} dx \tag{2.6}$$

$$w = -\int \int \frac{M}{EI} dx\, dx. \tag{2.7}$$

If E and I are constant along the beam then dw/dx and w can be found by repeated integration of the bending moment diagram. Figure 2.8(a) and (b) shows the diagrams of dw/dx and w for the beam of Fig. 2.4, obtained by repeated integration of the bending moment diagram of Fig. 2.4(c). Similar slope and deflection diagrams for other common design loads are given in Appendix A. If E and/or I vary along the beam, the bending moment diagram must be divided by EI before integration to obtain dw/dx and w.

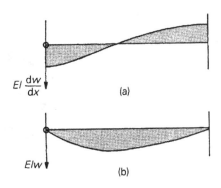

Fig. 2.8 (a) Slope and (b) deflection diagrams for load of Fig. 2.4.

Fig. 2.9 End forces and displacements on part of continuous beam.

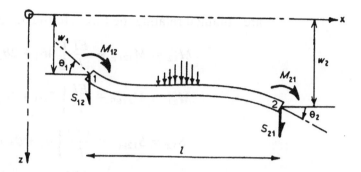

A beam which is relatively deep, or has small web area, can experience shear deformation which is significant in comparison to bending. The additional slope $(dw/dx)_S$ due to shear is related to the 'shear area' A_S of the beam and the shear modulus G by

$$\left(\frac{dw}{dx}\right)_S = \frac{S}{A_S G}.$$

Gere and Timoshenko [1] provide equations for the shear deformation of beams under various loadings. They use the term (A/α) in place of A_S above, where A is the cross-sectional area of the beam and α is the ratio of the shear stress at the centroid to the average shear stress S/A.

2.3.4 Analysis of continuous beams by slope–deflection equations

In contrast to the simply supported beam above, a continuous beam does not generally have zero moments at the supports. Figure 2.9 shows one span (or part of a span) of a continuous beam subjected to end moments M_{12}, M_{21} and shear forces S_{12} and S_{21} and with end displacements w_1, θ_1 and w_2, θ_2. In Fig. 2.9 and in equations (2.8) and (2.9) which follow, the end moments are positive clockwise and end shear forces positive downwards. This sign convention is adopted for these equations in preference to the sagging positive convention above so that moments or shear forces can be added in equilibrium equations without having to consider at which end of the beam they act.

By solving equations (2.6) and (2.7) for the beam of Fig. 2.9 with its end forces and displacements, the following slope–deflection equations

Fig. 2.10 Fixed end moments and shear forces.

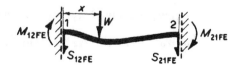

are obtained relating the forces to the deflections:

$$M_{12} = M_{12\text{FE}} + \frac{EI}{l}\left[4\theta_1 + 2\theta_2 + \frac{6}{l}(w_1 - w_2)\right]$$

$$M_{21} = M_{21\text{FE}} + \frac{EI}{l}\left[2\theta_1 + 4\theta_2 + \frac{6}{l}(w_1 - w_2)\right] \quad (2.8)$$

$$S_{12} = S_{12\text{FE}} + \frac{6EI}{l^2}\left[\theta_1 + \theta_2 + \frac{2}{l}(w_1 - w_2)\right]$$

$$S_{21} = S_{21\text{FE}} + \frac{6EI}{l^2}\left[\theta_1 + \theta_2 + \frac{2}{l}(w_1 - w_2)\right].$$

Without terms $M_{12\text{FE}}$, $M_{21\text{FE}}$, $S_{12\text{FE}}$ and $S_{21\text{FE}}$, equations (2.8) give the end forces in terms of the end deflections for a beam having no load along its length. If there are intermediate loads acting on the beam, their effect is introduced by the terms $M_{12\text{FE}}$, etc. which are identical to the forces that would act on the end of the beam if the ends were rigidly encastered with $w_1 = w_2 = \theta_1 = \theta_2 = 0$. Figure 2.10 shows a fixed end beam supporting a point load W at distance x from one end. For this beam the fixed end moments and shear forces are

$$M_{12\text{FE}} = -\frac{W(l-x)^2 x}{l^2}$$

$$M_{21\text{FE}} = \frac{Wx^2(l-x)}{l^2} \quad (2.9)$$

$$S_{12\text{FE}} = -\frac{W(l-x)^2(l+2x)}{l^3}$$

$$S_{21\text{FE}} = -\frac{Wx^2(3l-2x)}{l^3}.$$

The fixed end forces for distributed or varying loads can be obtained by integrating equations (2.9).

Equations (2.8) and (2.9) apply only to beams with constant EI. Similar equations can be written for tapered and haunched beams; useful charts are contained in Lightfoot [5]. However, in a computer analysis it is common for tapered beams to be considered as a number of connected prismatic lengths of reducing size, for each of which equations (2.8) and (2.9) can be derived.

By considering vertical and rotational equilibrium of any support of a continuous beam such as at 1 in Fig. 2.11, two equations are obtained:

$$a_{01}w_0 + a_{11}w_1 + a_{12}w_2 - b_{01}\theta_0 + b_{11}\theta_1 + b_{12}\theta_2 + S_{10\text{FE}} + S_{12\text{FE}} = 0$$

$$b_{01}w_0 + b_{11}w_1 - b_{12}w_2 + c_{01}\theta_0 + c_{11}\theta_1 + c_{12}\theta_2 + M_{10\text{FE}} + M_{12\text{FE}} = 0$$

$$(2.10)$$

Fig. 2.11 Two spans of continuous beam deck.

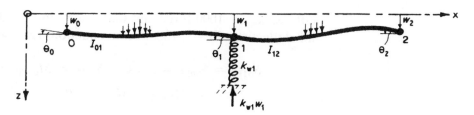

where

$$a_{01} = -\frac{12EI_{01}}{l_{01}^3} \qquad b_{01} = \frac{6EI_{01}}{l_{01}^2} \qquad c_{01} = \frac{2EI_{01}}{l_{01}}$$

$$a_{11} = -a_{01} - a_{12} + k_{w1} \qquad b_{11} = -b_{01} + b_{12} \qquad c_{11} = 2c_{01} + 2c_{12} + k_{\phi 1}$$

and k_{w1} and $k_{\phi 1}$ are vertical and rotational stiffnesses of support 1.

Similar pairs of stiffness equations can be written for every other support, so that for N supports there are $2N$ equations for $2N$ unknown deflections. These equations are solved to give the deflections, which can then be substituted back into equation (2.8) to give the moments and shear forces along the spans. As shown below, the number of equations and unknown deflections can be reduced if supports are rigidly restrained against either vertical or rotational movement.

The above stiffness equations provide a simple method of analysis of continuous beams using programmable desktop calculators. Such machines can usually solve a reasonable number of simultaneous equations. The method is demonstrated below with a worked example. This example can also be solved simply by hand using moment distribution, as described in Lightfoot [5].

(c) Worked example
Figure 2.12(a) shows a three-span bridge with piers 0 and 2 pinned for rotation and rigid vertically, 1 pinned for rotation and with vertical

Fig. 2.12 Moments in three-span beam.

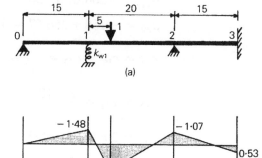

stiffness $k_{w1} = 1000$ force/unit deflection, and 3 rigid against rotation and deflection. $EI = 10\,000$. From equations (2.9) we find

$$S_{01FE} = S_{10FE} = S_{23FE} = S_{32FE} = M_{01FE} = M_{10FE}$$
$$= M_{23FE} = M_{32FE} = 0$$

$$S_{12FE} = -\frac{1(20-5)^2(20+2\times 5)}{20^3} = -0.844$$

$$S_{21FE} = -\frac{1\times 5^2(3\times 20 - 2\times 5)}{20^3} = -0.156$$

$$M_{12FE} = -\frac{1(20-5)^2 5}{20^2} = -2.81$$

$$M_{21FE} = \frac{1\times 5^2(20-5)}{20^2} = 0.94.$$

The stiffness coefficients are

$$a_{01} = a_{10} = a_{23} = a_{32} = -\frac{12\times 10\,000}{15^3} = -35.56$$

$$a_{12} = a_{21} = -\frac{12\times 10\,000}{20^3} = -15.0$$

$$a_{11} = 35.56 + 15.0 + 1000 = 1050.56$$

$$b_{01} = b_{10} = b_{23} = b_{32} = \frac{6\times 10\,000}{15^2} = 266.67$$

$$b_{12} = b_{21} = \frac{6\times 10\,000}{20^2} = 150.0$$

$$b_{00} = 0 + 266.67 = 266.67$$

$$b_{11} = -266.67 + 150 = -116.67$$

$$b_{22} = -150 + 266.67 = 116.67$$

$$c_{01} = c_{10} = c_{23} = c_{32} = \frac{2\times 10\,000}{15} = 1333$$

$$c_{12} = c_{21} = \frac{2\times 10\,000}{20} = 1000$$

$$c_{00} = 2\times 0 + 2\times 1333 = 2666$$

$$c_{11} = 2\times 1333 + 2\times 1000 = 4666$$

$$c_{22} = 2\times 1000 + 2\times 1333 = 4666$$

Since $w_0 = w_2 = w_3 = \phi_3 = 0$, we do not need to include these in stiffness equations and we can omit equations for corresponding vertical equilibrium at 0, 2 and 3 and rotational equilibrium at 3. Hence writing equations (2.10) for vertical equilibrium at 1, rotational equilibrium at 0, rotational equilibrium at 1 and rotational equilibrium at 2:

$$1050.56w_1 - 266.67\theta_0 - 116.67\theta_1 + 150\theta_2 \quad -0.844 = 0$$
$$-266.67w_1 + 2666\theta_0 + 1333\theta_1 \quad\quad\quad\quad\quad\quad\quad\quad = 0$$
$$-116.67w_1 + 1333\theta_0 + 4666\theta_1 + 1000\theta_2 \quad -2.81 \ = 0$$
$$150 \ w_1 \quad\quad\quad\quad + 1000\theta_1 + 4666\theta_2 + 0.94 \quad\quad = 0$$

which can be solved to give

$$w_1 = 0.00087 \quad \theta_0 = -0.000313 \quad \theta_1 = 0.000799 \quad \theta_2 = -0.00040$$

By substituting these back into equations (2.8) we obtain the moment diagram of Fig. 2.12(b).

2.3.5 Analysis of continuous beams by flexibility coefficients

The preceding stiffness equations were derived by assuming the structure has certain unknown deflections at supports and then deriving equilibrium equations for each support in turn in terms of the deflections. The equation for equilibrium of moments or vertical forces at any support can be thought of as

$$\Sigma [(\text{force on joint 1 due to unit deflection at 2}) \times (\text{deflection at 2})]$$
$$+ \text{ applied load on } 1 = 0.$$

where Σ is the sum for all deflections 1 to n, or

$$\Sigma [\text{stiffness coefficient} \times \text{deflection}] + \text{applied load} = 0.$$

An alternative approach, using 'flexibility' or 'influence' coefficients is sometimes more convenient and is demonstrated below. The indeterminate structure, such as the continuous beam in Fig. 2.13(a), is notionally 'released' in a number of places so that it becomes statically determinate. The problem is then to determine what equal and opposite actions X are to be applied to the two sides of every release so that all the releases close. To calculate the values of the indeterminate actions X, the opening of each release is calculated due to unit values of each action X in turn and due to the applied load. Then, for release at 1 to close, for example, the sum of the openings of the release due to all the loads is zero, which, expressed as an equation, is

$$\Sigma [(\text{opening of release 1 due to unit action } X_2) \times X_2]$$
$$+ \text{ opening of release 1 due to applied load} = 0$$

Fig. 2.13 Continuous beam = superposition of released beams.

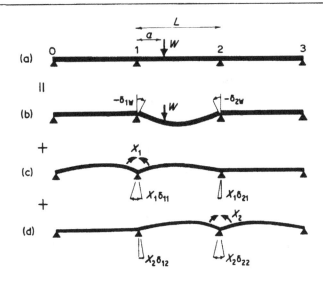

where Σ is the sum for all actions 1 to n

or

Σ [flexibility coefficients $\times$ action X] + opening due to applied load = 0.

By writing similar flexibility equations for all of the releases and solving for the actions X we obtain the values of the actions relevant for all the cuts to close simultaneously.

The above equations are expressed algebraically by

$$\begin{aligned} \delta_{11} X_1 + \delta_{12} X_2 + \cdots + \delta_{1n} X_n + \delta_{1W} &= 0 \\ \delta_{21} X_1 + \delta_{22} X_2 + \cdots + \delta_{2n} X_n + \delta_{2W} &= 0 \\ &\vdots \\ \delta_{n1} X_1 + \delta_{n2} X_2 + \cdots + \delta_{nn} X_n + \delta_{nW} &= 0 \end{aligned} \quad (2.11)$$

where

δ_{jk} = opening of cut j due to unit action X_k,

δ_{jW} = opening of cut j due to applied loads.

From considerations of flexural strain energy [6] it can be shown

$$\delta_{12} = \int \frac{m_1 m_2}{EI} \, dx$$

$$\delta_{1W} = \int \frac{m_1 m_W}{EI} \, dx \quad (2.12)$$

Fig. 2.14 Moments due to applied loads and unit release actions in Fig. 2.13.

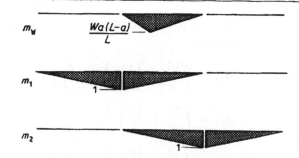

where, as illustrated in Fig. 2.14, m_1 and m_2 are moments in the released structure due to actions $X_1 = 1$ and $X_2 = 1$, respectively, and m_W are the moments in the released structure due to the applied loads. The values of the product integrals $\int mm\,dx$ for common moment diagrams are given in Appendix A, Fig. A.1. Their application is demonstrated below.

Worked example

Figure 2.15 shows a four-span bridge deck supporting a point load near the middle of the second span. Figure 2.15(b) shows the applied load moment diagram if the structure is released by relaxing the moments at 1, 2 and 3. Figure 2.15(c)–(e) shows the moment diagrams m_1, m_2 and

Fig. 2.15 Flexibility analysis of continuous beam: (a) load; (b) load moments; (c–e) unit release moments; (f) final moment diagram.

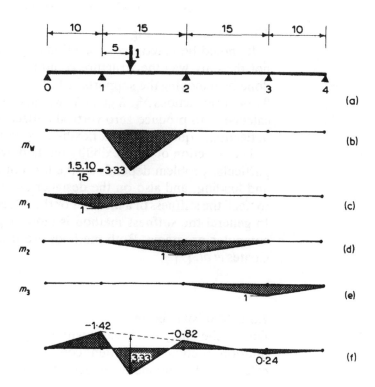

m_3 due to unit actions at 1, 2 and 3, respectively. From Appendix A we obtain:

$$\delta_{11} = \left(\frac{10}{3} + \frac{15}{3}\right)\frac{1}{EI} = \frac{8.33}{EI} \qquad \delta_{12} = \left(\frac{15}{6}\right)\frac{1}{EI} = \frac{2.5}{EI} \qquad \delta_{13} = 0$$

$$\delta_{22} = \left(\frac{15}{3} + \frac{15}{3}\right)\frac{1}{EI} = \frac{10}{EI} \qquad \delta_{23} = \left(\frac{15}{6}\right)\frac{1}{EI} = \frac{2.5}{EI}$$

$$\delta_{33} = \left(\frac{15}{3} + \frac{10}{3}\right)\frac{1}{EI} = \frac{8.33}{EI}$$

$$\delta_{1W} = \frac{15}{6} \times 3.33 \times 1 \times (1 + 0.66) \times \frac{1}{EI} = \frac{13.87}{EI}$$

$$\delta_{2W} = \frac{15}{6} \times 3.33 \times 1 \times (1 + 0.33) \times \frac{1}{EI} = \frac{11.1}{EI}.$$

Hence the flexibility equations can be written, (omitting $1/EI$)

$$8.33X_1 + 2.5X_2 \qquad\qquad + 13.87 = 0$$
$$2.5\ X_1 + 10\ X_2 + 2.5\ X_3 + 11.10 = 0$$
$$\qquad\qquad 2.5X_2 + 8.33X_3 + 0 \qquad = 0$$

which, when solved, give the unknown moments over the supports

$$X_1 = -1.42 \qquad X_2 = -0.82 \qquad X_3 = +0.24.$$

These are shown in Fig. 2.15(f).

It should be noted that relaxation of the moments at the support is not the only way the structure could have been released. It could be done by removing the supports at 1, 2 and 3 so that the deck spans from 0 to 4. Then actions X_1, X_2 and X_3 would be vertical forces and would be calculated to produce zero vertical deflection at the release points (or deflections equal to support flexibility × vertical reactions).

The selection of the flexibility method or the stiffness method for a particular problem depends on the form of the structure, its supports, and loading, and also on the designer as to whether he finds it easier to 'feel' the stiffness of a member or the effect of a release displacement. In general the stiffness method is more popular, and easier to use in computer programs. Both methods are described in greater detail in Coates *et al.* [7].

2.3.6 Multispan beams

During the analysis of beams with many approximately equal spans of uniform section it is often convenient to reduce the amount of calculation by notionally replacing the unloaded spans to the side of the

Fig. 2.16 Approximate analysis of multiple span beams. (a) prototype; (b) approximate analysis of reduced structure; and (c) moments in load-free regions.

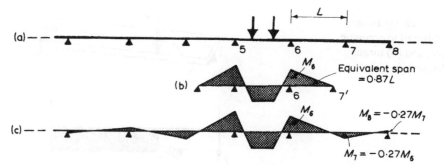

region of consideration by a single span of approximately 0.87 times its real length, as shown in Fig. 2.16(b). The moment calculated for the penultimate support in the reduced structure will be close to that in the extended structure. Having determined this moment, those in the omitted internal supports are found simply by progressing from the loaded region and making each support moment − 0.27 of the preceding support moment as shown in Fig. 2.16(c).

2.4 TORSION OF BEAMS

2.4.1 Equilibrium of torques

In Fig. 2.3 the vertical load dW was acting above the centre line of the beam and was held in equilibrium solely by the shear forces S and moments M. If the vertical load dW is placed eccentric to the centre line as in Fig. 2.17 then additional actions in the form of torques T and $T + dT$ on the ends of the element are necessary to retain equilibrium of couples about the longitudinal axis Ox. On taking moments about Ox we obtain

$$dT = y\,dW \qquad (2.13)$$

where y is the eccentricity of the load from the longitudinal centre line. If the beam does not have a vertical axis of symmetry then y should be

Fig. 2.17 Torque on elememt of beam.

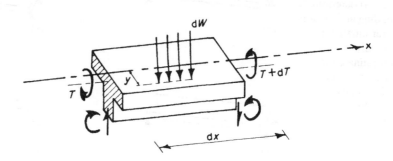

Fig. 2.18 (a) Beam subjected to eccentric load and (b) torque diagram.

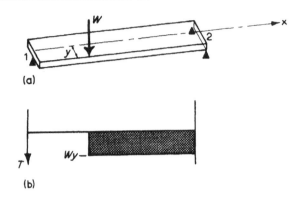

measured from the shear centre [1–4]. Since most beam decks are symmetrical, this added complexity is not dealt with further here.

Figure 2.18(a) shows a simply supported beam deck, with two bearings at end 2 and one bearing at end 1. Under the action of the eccentric load W near midspan the beam is subjected to a torque. With only one bearing at end 1, no torque can be transmitted to the support. To determine the torque T on any cross-section, moments can be taken about Ox for the beam to the left of the section. In this way, the torque diagram in Fig. 2.18(b) is obtained.

It should be noticed that a beam is only statically determinate for torsion when it has only one pair of bearings to resist torques. Figure 2.19 shows a number of different bearing arrangements for beams which are: statically determinate in both bending and torsion, statically determinate in torsion but not bending, statically determinate in bending but not torsion, and statically indeterminate in both bending and torsion. If all pairs of bearings are placed at right angles to the longitudinal axis then the equilibrium (and analysis) of the beam in bending and torsion can be considered separately. However if a pair of bearings are skew at a pier, the moments and torques interact at the pier

Fig. 2.19 Statical determinacy of beam in torsion: (a) determinate for bending and torsion; (b) determinate for torsion only; (c) determinate for bending only; and (d) indeterminate for both bending and torsion.

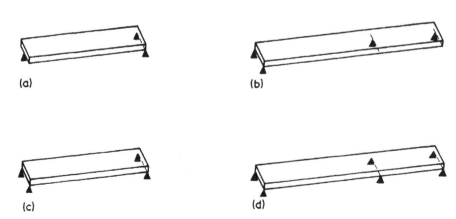

Torsion of beams

and the analysis is more complicated and is best performed with a two-dimensional grillage as described in Chapter 3. An exception is the simple case of a beam supported on only three bearings, not in line, when all the reactions can always be determined by taking moments about two axes.

2.4.2 Torque–twist relationship

To determine the distribution of torques in a beam with four bearings, as in Fig. 2.20, some knowledge of its deformation behaviour is needed. It is only if the beam has uniform torsional stiffness that simple physical argument leads to the solution.

When a torque T is applied to an element of beam it causes the element to twist about axis Ox with a relative rotation $d\phi$ between the ends of the element. For an elastic material, the amount of relative rotation is proportional to the torque and related by the equation

$$T = -CG \frac{d\phi}{dx} \tag{2.14}$$

where C is the torsion constant of the section, discussed below. G is the shear modulus $= E/2(1 + v)$ where v is Poisson's ratio. From equation (2.14)

$$d\phi = -\frac{T}{CG} dx$$

$$\phi = -\int \frac{T}{CG} dx. \tag{2.15}$$

The distribution of torque and rotation along a beam of uniform torsional rigidity CG restrained at both ends against rotation can be

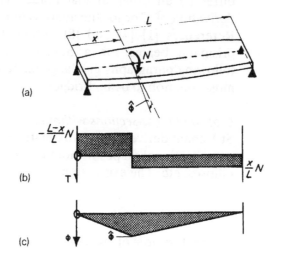

Fig. 2.20 Torsion of beam with both ends restrained: (a) loading; (b) torque diagram; and (c) rotation diagram.

found as follows. Figure 2.20 shows such a beam supporting a concentrated applied torque N at some point along the span. Since there are no other loads between N and the ends the torques must be constant (though different) along these lengths, as shown in Fig. 2.20(b). The rotation ϕ varies along the beam from zero at a support to ϕ at the load and back to zero at the other support. Since the torques are constant on each side of the load so must $d\phi/dx$ be, from equation (2.14). Hence the ϕ diagram in Fig. 2.20(c) must be triangular, and $d\phi/dx$ on each side must be inversely proportional to the distance of the load from each end. Consequently, the applied torque N is distributed to the two supports in fractions inversely proportional to the distance of the load from the support. It is thus found for this concentrated applied torque that the distribution of torque is similar to the distribution of shear force due to concentrated vertical load and the distribution of ϕ similar to the distribution of moment. Since a distributed load can be thought of as a superposition of concentrated loads, it is possible to derive the distribution of T and ϕ for any load on a uniform beam between two torque-resisting supports by direct analogy with shear force and moment distributions for a simple span. It should be noted that this analogy does not hold if CG is not constant.

2.4.3 Torsion constant C

The torsion constant C (often referred to as the St Venant torsion constant) is not generally a simple geometrical property of the cross-section in the same way that the flexural constant I is the second moment of area. In the case of a cylinder, C is identical to the polar moment of inertia I_p. However, this is a special case which can be misleading since, for many shapes of cross-sections, C is totally different from I_p, and can differ by an order of magnitude. There is no general rule for the derivation of C or for the analysis of torsional shear stress distribution. References [1]–[4] and [8] outline the elastic theory of torsion of prismatic beams for a number of shapes of cross-section. The following paragraphs give approximate rules appropriate to the shape most common to beam bridges.

C of solid cross-sections without re-entrant corners
St Venant derived an approximate expression which is applicable to all cross-sectional shapes without re-entrant corners, i.e. triangles, circles, ellipses, etc. The expression is

$$C = \frac{A^4}{40 I_p} \tag{2.16}$$

where A = area of cross-section and I_p = polar moment of inertia.

Fig. 2.21 Solid rectangular cross-section.

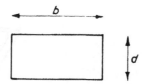

For a rectangle of sides b and d shown in Fig. 2.21, this can be reduced to

$$C = \frac{3b^3 d^3}{10(b^2 + d^2)} \qquad (2.17)$$

and, in the case of a thin rectangle $b > 5d$, this is more accurately given by

$$C = \frac{bd^3}{3}. \qquad (2.18)$$

The maximum shear stress on a rectangular cross-section occurs at the middle of the long side and has magnitude

$$\hat{\tau} = \frac{T}{bd^2(0.333 - 0.125\sqrt{d/b})} \qquad b > d. \qquad (2.19)$$

C of solid cross-sections with re-entrant corners
If the cross-section has re-entrant corners, C is very much less than that given by equation (2.16). Appendix B explains how the torsion constant C can be calculated for complex cross-sections with a finite difference method. However for many sections C can be obtained with sufficient accuracy by notionally subdividing the cross-section, as in Fig. 2.22, into shapes without re-entrant corners and summing the values of C for these elements. While doing this it is worth remembering Prandtl's membrane analogy, described by Timoshenko and Goodier [8]. It is shown that the stiffness of a cross-sectional shape is proportional to the volume under an inflated bubble stretched across a hole of the same shape. The shear stress at any point is along the direction of the bubble's contours and of magnitude proportional to the gradient at right angles to the contours. If

Fig. 2.22 Subdivision of section with re-entrant corners.

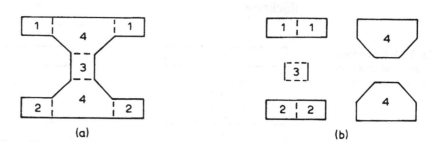

Fig. 2.23 Inflated membrane on cut rectangular cross-section.

a cross-section is cut in half, the membrane is in effect held down along the cut as in Fig. 2.23, thus greatly reducing its volume and preventing flow of shear stress along the contours from one part to the other. Consequently, when a section such as Fig. 2.22 is notionally split into elements it is important to choose the elements so that they maximize the volume under their bubbles. To avoid giving the bubble zero height at the notional cuts, the elements can be rejoined at the cuts for calculation as in Fig. 2.22(b). Since shear stresses flow across both ends of the web it can be thought of as part of a long thin rectangle for which $C = bd^3/3$. Figure 2.24 also shows the cross-section arbitrarily cut into rectangles which, by not trying to maximize the volume under the bubble, leads to a value of C of only half the correct figure.

C of thick-walled hollow sections
The torsion constant of a thick-walled hollow section can be found with sufficient accuracy for most practical purposes by calculating C for the shape of the outside boundary and deducting the value of C calculated for the inside boundary.

C of thin-walled hollow sections
The torsion constant of a thin-walled hollow section, such as in Fig. 2.25, is given by

$$C = \frac{4A^2}{\oint \frac{ds}{t}} \qquad (2.20)$$

where A is the area enclosed by the centre line of the walls and $\oint ds/t$ is the integral round the wall centre line of the length divided by the wall thickness.

Fig. 2.24 Erroneous subdivision of section with re-entrant corners.

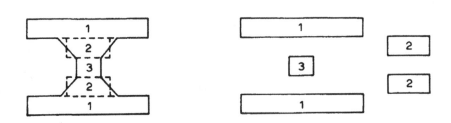

Fig. 2.25 Thin-walled box section.

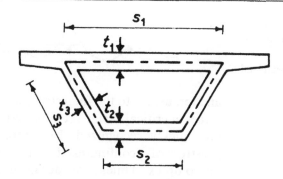

The shear stress round the wall at any point is given by

$$\tau = \frac{T}{2At}. \tag{2.21}$$

Equation (2.20) is only really applicable for cross-sections with only one cell or two symmetric cells. The equation can be applied to the outside boundary of a multicellular box as a first approximation, but for a more accurate calculation see Timoshenko [2] and Timoshenko and Goodier [8]. Chapter 6 explains how thin-walled hollow sections do not exhibit the torsional stiffness of equation (2.20) if the sections do not have internal bracing to hold them in shape.

2.5 COMPUTER ANALYSIS OF CONTINUOUS BEAMS

There are numerous different computer programs for analysis of continuous bridges. In general they are simple to use, and the derivation of computer input section properties is straightforward. It is not possible to mention all the facilities that are available, but convenient programs exist for the analysis of beams which are simply supported, continuous, prismatic, varying in section, supported or restrained by elastic supports. The analysis of a large number of load cases including prestress, settlement and temperature can be requested by few instructions. Output can be in terms of bending moment and shear force distributions, displacements, influence lines, or envelopes of maximum and minimum moments and shear forces along the deck. Not all these facilities are necessarily available in a single program; in fact as a general rule, the more versatile a program is, the more difficult it is to use. It is thus worth choosing the program to suit the complexity of the problem, unless a user has sufficient familiarity with one program in particular to enable him to adapt his use of it as necessary.

2.6 CONSTRUCTION SEQUENCE

The analysis of live load actions is generally a straightforward procedure of applying a number of different load cases with various dispositions

Fig. 2.26 Moments in continuous beam due to simultaneous application of dead load from end to end.

and intensities to the final structure, and inspecting moments and shear forces at critical design sections. With experience, the critical load cases are quickly recognized. In contrast, the analysis of dead load and prestress of a continuous deck built in a number of different stages can be complex. Unless the design is very conservative, the method of construction must be considered during design and, vice versa, the method of design must be remembered during construction.

If a structure is built from end to end on falsework prior to removal of the falsework, then the dead load moments should be as calculated for simultaneous application of the dead load to continuous beam, as shown in Fig. 2.26. On the other hand, if the bridge is erected span by span, as were the boxes of Stephenson's Britannia Bridge, simple moment connections of the spans over the supports do not of themselves induce large dead load moments over the supports which reduce span moments. These moments can be induced only by removing the relative rotations due to self weight of the touching ends of adjacent spans. This can be done by jacking a rotation at the joint, or by bolting up with prestressing bolts, or, as Stephenson did, by connecting spans as in Fig. 2.27 with one end temporarily raised. The articulation and moments of his boxes changed as each span was added, and each had to be designed as simply supported for dead load during construction and continuous for dead load and live load in use.

Multiple span cast *in situ* concrete bridges are often built span by span or two spans at a time with cantilevers as in Fig. 2.28. The simplest way of visualizing the dead loading due to any stage of construction is to consider the structure just before and after the falsework is struck. The load added to the continuous deck of cured concrete is identical to the load previously carried by the falsework. The bending moment diagram in Fig. 2.28 with alternate high and low moments at supports is typical of two-span by two-span construction. If the deck is prestressed in the

Fig. 2.27 Induction of internal support moments by lowering ends.

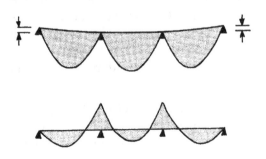

Fig. 2.28 Two-span by two-span construction.

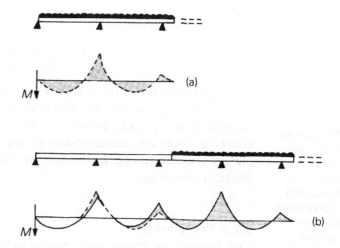

same stages, the falsework is effectively struck when the prestress picks up the dead load (except near supports where falsework will carry part of the reaction until its elastic compression is released). If the prestress profile is the same in every span it also generates support moments of alternating magnitude, which, being opposite in sign to dead load moments, virtually cancel out the effects of stage by stage construction. The analysis of prestress moments is discussed in Chapter 11.

2.7 FRAME AND ARCHING ACTION

2.7.1 Axial force with bending

Bridges which take advantage of the frame action and arching geometry are often able to achieve greater slenderness and elegance than beam bridges. The principles of computer analysis are the same as for beams except that axial compression is included in the stiffness analysis and detailed attention has to be given to the horizontal stiffness as well as the vertical stiffness of the foundations.

Frame and arching structures are most easily modelled in a plane frame, or space frame, using straight members. The bending moments along each member are then independent of the axial load, which is not the case if a curved member is used. The stresses at any point are the combination of bending stress given by equation (2.3) and the axial stress

$$\sigma = \frac{P}{A} \qquad (2.22)$$

where P is the axial force on the section and A the cross-sectional area.

48   Beam decks and frames

The combined stresses are

$$\sigma = \frac{P}{A} \pm \frac{Mz}{I}. \tag{2.23}$$

The sign conventions for P and M are not standardized since different conventions are useful in different situations. On steel structures tension is often made positive with compression negative, while on concrete structures compression is usually considered positive. It is always sensible to check the signs of axial and bending stresses using physical reasoning.

2.7.2 Portal bridge

The benefits of frame and arching action are demonstrated by the diagrams of portal bridge structures in Fig. 2.29. Figure 2.29(a) shows

Fig. 2.29 Portal bridge moments: (a) simply supported deck; (b) portal with foundations assumed pinned on rigid ground; (c) foundations on deforming ground; (d) with haunches; (e) and (f) lines of thrust.

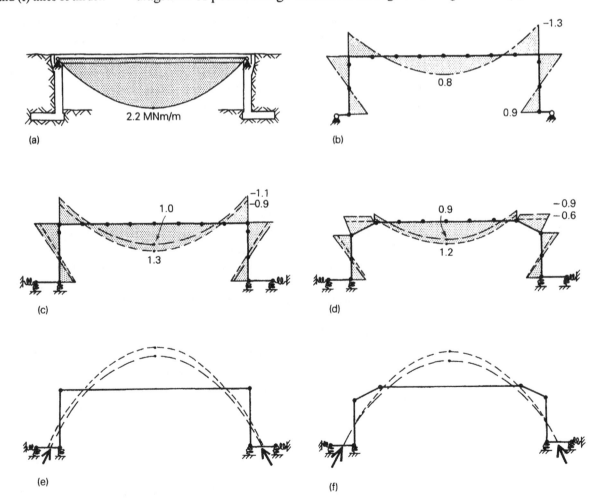

the bending moment diagram for a simply supported span of 24 m, without any frame action, supporting a uniformly distributed load of 0.030 MN/m^2. Figure 2.29(b) shows the bending moments for the same span when it forms part of a portal frame of uniform section with height 7 m from foundations which are assumed pinned.

The support and restraint conditions provided by the ground need careful assessment and different assumptions can lead to markedly different results. Chapter 14 presents equations for the stiffness of foundations and information on soil properties. Particular care has to be taken if the embankments are likely to move due to subsidence or settlement of substrata since large forces may develop due to the weight of moving soil. However, in many situations the uncertainties relating to soil stiffnesses are not markedly greater than the uncertainties relating to the structure stiffnesses, and reasonable estimates can be made of upper and lower conditions.

Figure 2.29(c) indicates upper and lower estimates of the bending moment diagram for the portal of Fig. 2.29(b) with elastic foundations. The diagram with the smaller moment of 1.0 MNm at midspan was calculated assuming short-term effective moduli of 100 MPa for ground of very stiff clay and 30 000 MPa for the concrete structure. The other moment diagram with moment 1.3 MNm at midspan was calculated with Young's modulus for the ground reduced by a factor of 3 to correspond approximately to drained ground conditions, as explained in Chapter 14, with the concrete modulus still 30 000 MPa. If the concrete structure creeps over a period of time, so that its effective modulus becomes 10 000 MPa (as discussed in Section 2.8) the bending moment at midspan may creep back towards the smaller value of 1 MNm.

Figure 2.29(d) shows bending moments in a similar portal with haunches, calculated with the same assumptions as Fig. 2.29(c). Fig. 2.29(e) and (f) illustrate the forces of Fig. 2.29(c) and (d) in terms of lines of thrust. The bending moment at any point is equal to the horizontal component of the thrust times the vertical distance from the point to the line of thrust. Since the horizontal component is constant across the bridge the bending moment diagram for each analysis is similar in shape to the vertical offset of the line of thrust from the structure.

2.7.3 Arching bridge

A further example of arching action is shown in Fig. 2.30 relating to a three-span footbridge with spans of 25 m, 35 m and 25 m. In Fig. 2.30(a) the deck is treated as a continuous beam supporting a uniformly distributed load of 0.040 MN/m. In Fig. 2.30(b) the deck is supported on

Fig. 2.30 Arching bridge moments:
(a) deck treated as continuous beam;
(b) deck arching on raking piers on rigid ground; and
(c) foundations on deforming ground.

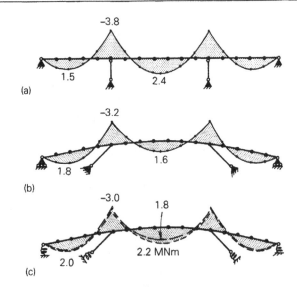

raking piers, with the foundations behaving as pinned on hard ground (or tied by bars under the road). Figure 2.30(c) shows the effects of lateral spread of the foundations. The upper line with midspan moment of 1.8 MNm relates to concrete with Young's modulus of 30 000 MPa and stiff clay ground with Young's modulus of 40 MPa. The lower line with midspan moment 2.2 MNm relates to the same modulus for the structure with the ground softened by a factor of 3 to reflect drained conditions.

Bridges which utilize arching action can be more efficient structures than beams on columns when they are founded on very stiff ground. However, they can be much more difficult to build, particularly if the deck is post-tensioned. The additional costs due to construction complexity can in some circumstances be much larger than the economies from structural efficiency.

2.8 SHORT-TERM AND LONG-TERM BEHAVIOUR

The load distribution within indeterminate structures can change markedly at different times after construction. Construction materials creep over a period of time and foundations settle and give. The deformations of soils are discussed in Chapter 14. Creep of concrete, under a long-term load from early life, causes creep deformations of the same magnitude as initial elastic values after about six months to a year, and the same again after many years. Thus the final deflections are about three times the initial values. Detailed time-related calculations can be very complicated, and subject to significant uncertainty, particularly if the construction sequence is complex and if loads change with time. References [9]–[11] give guidance. Under long-term loading

Short-term and long-term behaviour 51

the structure slowly creeps towards the distributions that would have existed if it had been built all at once with structure and ground stiffnesses corresponding to long-term values. Peculiarities of deflected shape and stress due to the particular construction sequence are progressively masked by the long-term deflections.

An approximate estimate of the change due to creep can often be obtained by calculating the short-term behaviour for conditions as built and the long-term behaviour for the conditions of built-in-one. The long-term built-in-one conditions can be analysed approximately using an effective modulus of $E/(1 + \phi)$, where E is short-term Young's modulus and ϕ is the creep factor (ratio of creep deformation to initial elastic deformation). The effective modulus method has shortcomings for detailed analysis of creep when loadings change with time, as explained in Clark [9]. In the long term, conditions will tend towards a many other uncertainties and the effective modulus method is often adequate. A more accurate estimate of long-term changes can be obtained by considering the exponential decay of creep deformation, as explained in Clark [9]. In the long-term, conditions will tend towards a

Fig. 2.31 Steel box girder footbridge across the M5 motorway near Taunton, England. The raking piers are connected by tie bars under the carriageways. Designed by Somerset Sub Unit of South West Road Construction Unit. Photograph E.C. Hambly.

situation at the proportion $(1 - e^{-\varphi})$ of the difference between the short-term as-built and long-term built-in-one conditions. This proportion is about 0.6 after six months ($\phi = 1$) and 0.9 after many years ($\phi = 2$).

REFERENCES

1. Gere, J.M. and Timoshenko, S.P. (1987) *Mechanics of Materials*, Van Nostrand Reinhold, London, 2nd edn.
2. Timoshenko, S.P. (1955) *Strength of Materials*, 2 volumes, Van Nostrand Reinhold, New York, 3rd edn.
3. Roark, R.J. and Young, W.C. (1989) *Formulas for Stress and Strain*, McGraw-Hill, New York, 6th edn.
4. Oden, J.T. (1967) *Mechanics of Elastic Structures*, McGraw-Hill, New York.
5. Lightfoot, E. (1961) *Moment Distribution*, E. & F.N. Spon, London.
6. Morice, P.B. (1959) *Linear Structural Analysis*, Thames and Hudson, London.
7. Coates, R.C., Coutie, M.G. and Kong, F.K. (1988) *Structural Analysis*, Van Nostrand Reinhold, Wokingham, 3rd edn.
8. Timoshenko, S.P. and Goodier, J.N. (1970) *Theory of Elasticity*, McGraw-Hill, New York, 3rd edn.
9. Clark, L.A. (1983) *Concrete Bridge Design to BS5400*, Construction Press, Longmans, London, with supplement (1985).
10. Kong, F.K., Evans, R.H., Cohen, E. and Roll, F. (1983) *Handbook of Structural Concrete*, Pitman, London.
11. *CEB-FIP Model Code for Concrete Structures* (1978) Comité Euro-International du Béton.

3

Slab decks

3.1 INTRODUCTION

A slab deck is structurally continuous in the two dimensions of the plane of the slab so that an applied load is supported by two-dimensional distributions of shear forces, moments and torques. These distributions are considerably more complex than those along a one-dimensional continuous beam. This chapter presents the fundamental relationships of equilibrium and stress–strain behaviour of an element of a slab. Because rigorous solution of the basic equations for a real deck is seldom possible, a much used approximate method is described. This is grillage analysis in which the deck is represented for the purpose of analysis by a two-dimensional grillage of beams. Another approximate method called 'finite element analysis' is described in Chapter 13. In this method the deck is notionally subdivided into a large number of small elements for each of which approximate plate bending equations can be written and the whole set solved. Hand methods of analysis employing charts are also frequently used for slab decks with simple plan geometry. These methods are reviewed in Chapter 10. However, the steady improvements made to grillage programs in recent years now make this computer method more versatile, as quick as, and simpler to comprehend than chart methods.

3.2 TYPES OF STRUCTURE

Figure 3.1 shows some common forms of slab deck construction. In Fig. 3.1(a) the slab is of solid reinforced concrete. In (b) the weight has been reduced by casting voids within the depth of the slab, and the deck is referred to as a 'voided slab'. If the depth of the voids exceeds 60% of the depth of the slab, the slab may not behave like a single plate but more like a cellular deck for which analysis is described in Chapter 5. A slab deck can be constructed of composite construction as in Fig. 3.1(c) and (d). In (c) the slab has been constructed by casting in-fill concrete between contiguous beams with continuous transverse reinforcement top and bottom. In (d) the deck is constructed of contiguous box beams post-tensioned transversely to give moment continuity.

The slab decks in Fig. 3.1 can have similar stiffnesses in longitudinal and transverse directions in which case they are called 'isotropic'. If the stiffnesses differ in the two directions, as is likely for the decks of (c) and (d), then the slab is called 'orthotropic'.

Slab decks sometimes have their parapet edges stiffened by upstand and downstand beams as in Fig. 3.1(a). The deck is then equally able to carry a load in the centre with load distributed to the slab on both sides, and to carry the load near one edge with distribution to the slab on one side and stiffening beam on the other. While such edge stiffening presents problems for rigorous analysis, it does not complicate the approximate methods described later unless the upstand or downstand is so deep that the neutral axis locally is at a significantly different level from the midplane of the slab. The effects of such upstand beams and of service troughs are discussed in Chapter 8.

Bridges are frequently designed with their decks skew to the supports, tapered, or curved in plan. The behaviour and rigorous analysis is significantly complicated by the shape, but as shown in

Fig. 3.1 Slab decks: (a) solid; (b) voided; (c) composite solid; and (d) composite voided.

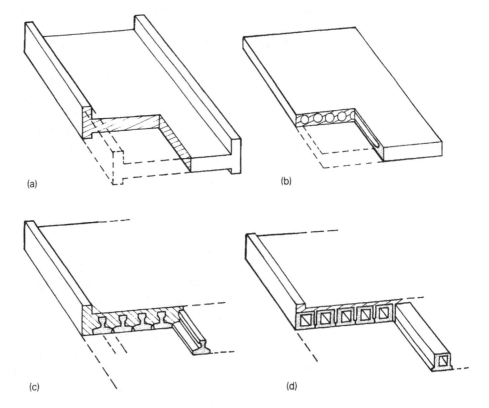

Chapter 9 the effect on grillage analysis is one of inconvenience rather than theoretical complexity.

3.3 STRUCTURAL ACTION

3.3.1 Equilibrium of forces

Figure 3.2 shows an element of slab subjected to vertical load dW and internal moments m, shear forces s and torques t (all per unit width) which interact with adjoining parts of the slab. By writing

$$\frac{\partial m_y}{\partial y} dy = dm_y, \quad \frac{\partial s_y}{\partial y} dy = ds_y \quad \text{etc, and} \quad W dx\, dy = dW$$

we obtain, on resolving vertically and taking moments about axes Ox and Oy, after simplification

$$\frac{\partial s_x}{\partial x} + \frac{\partial s_y}{\partial y} = -W \tag{3.1}$$

$$\frac{\partial m_x}{\partial x} + \frac{\partial t_{yx}}{\partial y} = s_x$$

$$\frac{\partial m_y}{\partial y} + \frac{\partial t_{xy}}{\partial x} = s_y. \tag{3.2}$$

These equations differ significantly from those for a single beam. In addition to the obvious difference that loads distribute in two

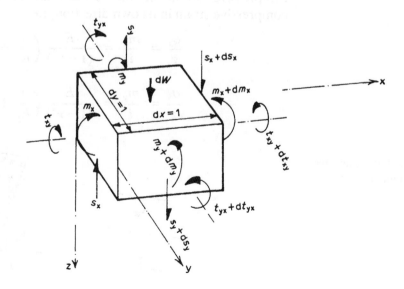

Fig. 3.2 Resultant forces on element of slab.

dimensions, equations (3.2) indicate that shear force is not the simple differential of the bending moment (i.e. it is not the slope of the bending moment diagram).

In grillage analysis the different components $\partial m_x/\partial x$ due to bending and $\partial t_{yx}/\partial y$ due to torsion exhibit themselves in different ways, and it is convenient to define

$$s_{Mx} = \frac{\partial m_x}{\partial x} \text{ and } s_{Tx} = \frac{\partial t_{yx}}{\partial y} \qquad (3.3)$$

so that equation (3.2) becomes

$$s_x = s_{Mx} + s_{Tx}.$$

At any level of the slab element the horizontal shear stresses on faces normal to Ox and Oy must be complementary to maintain equilibrium. Consequently, the torques on orthogonal faces of the slab element are also complementary and equal:

$$t_{xy} = t_{yx}. \qquad (3.4)$$

3.3.2 Moment–curvature equations

The simple theory of elastic bending of slabs is based on the same assumptions as simple beam theory. Lines in the slab normal to the neutral plane remain straight so that strains and bending stresses, shown in Fig. 3.3, increase linearly with distance from the neutral axis. Also, vertical compressive stresses are zero. However, unlike a simple beam, the compressive bending stress σ in one direction is dependent on the compressive strain in the orthogonal direction as well as the compressive strain in its own direction, i.e.

$$\frac{\sigma_x}{z} = \frac{m_x}{i} = \frac{E}{(1-v^2)}\left(\frac{1}{R_x} + \frac{v}{R_y}\right)$$

$$\frac{\sigma_y}{z} = \frac{m_y}{i} = \frac{E}{(1-v^2)}\left(\frac{1}{R_y} + \frac{v}{R_x}\right) \qquad (3.5)$$

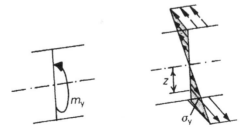

Fig. 3.3 Bending stress distribution.

or

$$m_x = -D\left(\frac{\partial^2 w}{\partial x^2} + \nu \frac{\partial^2 w}{\partial y^2}\right)$$

$$m_y = -D\left(\frac{\partial^2 w}{\partial y^2} + \nu \frac{\partial^2 w}{\partial x^2}\right)$$

where

$$D = \frac{Ei}{(1-\nu^2)} = \frac{Ed^3}{(1-\nu^2)12} \quad \text{is the flexural rigidity}$$

z = vertical distance of point below neutral axis

$i = d^3/12$ = second moment of area of slab per unit width

d = slab thickness

R_x = radius of bending curvature in the x direction

E = Young's modulus

ν = Poisson's ratio.

The torsion shear stresses in an element of slab have a linear distribution as shown in Fig. 3.4 with stress τ proportional to distance z from the neutral axis, so that

$$\frac{\tau_{xy}}{z} = \frac{t_{xy}}{i} = -\frac{E}{(1+\nu)}\left(\frac{1}{R_{xy}}\right)$$

$$t_{xy} = -\frac{Ed^3}{(1+\nu)12}\left(\frac{\partial^2 w}{\partial x \partial y}\right) = -\frac{Gd^3}{6}\left(\frac{\partial^2 w}{\partial x \partial y}\right)$$

(3.6)

where

$$G = \frac{E}{2(1+\nu)} = \text{elastic shear modulus.}$$

Equation (3.6) for t_{xy} can be written

$$t_{xy} = -cG \frac{\partial^2 w}{\partial x \partial y} \tag{3.7}$$

Fig. 3.4 Torsion stress distribution.

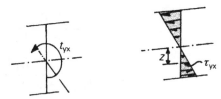

where c is the effective torsion constant per unit width of slab given by

$$c = \frac{d^3}{6} \text{ per unit width.} \qquad (3.8)$$

Equation (3.8) for the torsion constant of a slab per unit width is equal to half that in equation (2.18) for a thin slab-like beam. This difference is the consequence of a difference in definition of torque. If the twisted thin slab-like beam in Fig. 3.5 is analysed as a beam as in Section 2.4, then the torque T is defined as the sum of the torque due to the opposed horizontal shear flows near the top and bottom faces and of the torque due to the opposed vertical shear flows near the two edges. In contrast, if the slab-like beam of Fig. 3.5 is analysed as a slab, then the torque t_{xy} is defined as only due to the opposed horizontal shear flows near the top and bottom faces. The vertical shear flows at the edges constitute local high values of the vertical shear force s_x. The opposed vertical shear flows provide half the total torque and are associated by Equation (3.2) with the transverse torque t_{yx} defined in Fig. 3.2. The two definitions of torque, though different, are equivalent: while the slab has half the torsion constant (and hence half strain energy) of the 'beam' attributed to longitudinal torsion, the other half of the torsion constant (and strain energy) is attributed to transverse torsion not considered in beam analysis.

Equations (3.5)–(3.8) relate to isotropic slabs whose elastic behaviour can be described by the constants E and v. If the slab is orthotropic, Young's modulus and Poisson's ratio are different in the two directions and the moment–curvature equations are much more complicated.

$$m_x = -D_x \left(\frac{\partial^2 w}{\partial x^2} + v_y \frac{\partial^2 w}{\partial y^2} \right)$$

$$m_y = -D_y \left(\frac{\partial^2 w}{\partial y^2} + v_x \frac{\partial^2 w}{\partial x^2} \right)$$

$$t_{xy} = -2D_{xy} \left(\frac{\partial^2 w}{\partial x \, \partial y} \right)$$

where the flexural rigidities are

$$D_x = \frac{E_x d^3}{(1 - v_x v_y)12}$$

$$D_y = \frac{E_y d^3}{(1 - v_x v_y)12}$$

$$D_{xy} = \frac{E_x E_y d^3}{[E_x(1 + v_{yx}) + E_y(1 + v_{xy})]12} = \frac{G_{xy} d}{12}$$

$$\simeq \tfrac{1}{2}(1 - v_x v_y) \sqrt{(D_x D_y)}.$$

(3.9)

Fig. 3.5 Torsion of slab-like beam.

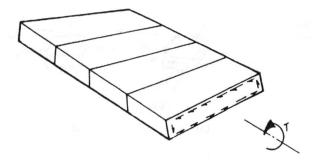

The expression for D_{xy} is an approximation determined by Huber [1, 2].

3.3.3 Principal bending moments and principal stresses

The element of slab in Fig. 3.6(a) has been defined with faces normal to axes Ox and Oy. The faces are subjected to combinations of moment and torsion $m_x, t_{xy}; m_y, t_{yx}$. If the element is defined with faces normal to axes in other directions, the magnitudes of the moments and torques are different. With axes in one particular set of directions, called the principal directions, the torques disappear as in Fig. 3.6(b) and the moments m_I and m_{II} on the faces represent the maximum and minimum moments at that point in the slab. If α is the angle between the axis Ox in Fig. 3.6(a) and the axis I–I in Fig. 3.6(b), the principal moments m_I and m_{II} are related to m_x, m_y and t_{xy} by the equation

$$m_I = \frac{m_x + m_y}{2} + \sqrt{\left[\left(\frac{m_x - m_y}{2}\right)^2 + t_{xy}^2\right]}$$

$$m_{II} = \frac{m_x + m_y}{2} - \sqrt{\left[\left(\frac{m_x - m_y}{2}\right)^2 + t_{xy}^2\right]} \quad (3.10)$$

$$\tan 2\alpha = \frac{2t_{xy}}{m_x - m_y}.$$

This relationship between moments and torques on faces normal to various axes can be represented by Mohr's circle shown in Fig. 3.6(c).

Fig. 3.6 Principal moments and Mohr's circle of moment.

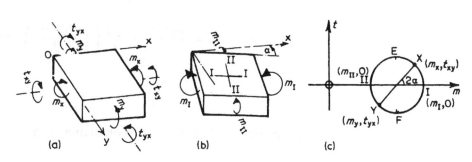

Fig. 3.7 Principal stresses and Mohr's circle of stress.

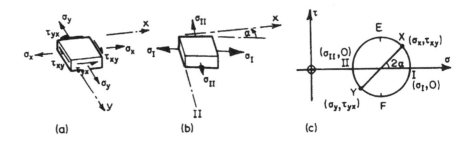

With axes for moment m and torque t (here defined as positive for right-handed screw torque vectors towards the centre of the element, so that $t_{yx} = -t_{xy}$) the circle is drawn with points (m_x, t_{xy}) and (m_y, t_{yx}) at opposite ends of a diameter. The points $(m_I, 0)$ and $(m_{II}, 0)$ also on opposite ends of a diameter represent the major and minor principal moments. It should be noted that a difference in direction of axes of α in the element is represented by a difference in inclination of diameters of 2α on Mohr's circle. The maximum torque occurs on the circle at points E and F at which moments are equal and which are at opposite ends of the diameter at 90° to $m_I m_{II}$. In the element, the maximum torques occur on faces normal to axes at 45° to principal axes. From the geometry of the circle it is evident that the maximum torque at E or F is

$$t = \frac{m_I - m_{II}}{2} = \sqrt{\left[\left(\frac{m_x - m_y}{2}\right)^2 + t_{xy}^2\right]}.$$

Tensile stresses and shear stresses on orthogonal planes through a point in the slab, as in Fig. 3.7, are related by the same rules of equilibrium as moments and torques so that Mohr's circle is also used. The sign convention for shear stresses is positive stresses are in a direction clockwise round the element, so that $\tau_{yx} = -\tau_{xy}$. Consequently the compressive and shear stresses σ_x, σ_y, τ_{xy} on faces normal to axes Ox and Oy and the principal stresses σ_I, σ_{II} with axes inclined at α to Oxy are related by

$$\sigma_I = \frac{\sigma_x + \sigma_y}{2} + \sqrt{\left[\left(\frac{\sigma_x - \sigma_y}{2}\right)^2 + \tau_{xy}^2\right]}$$

$$\sigma_{II} = \frac{\sigma_x + \sigma_y}{2} - \sqrt{\left[\left(\frac{\sigma_x - \sigma_y}{2}\right)^2 + \tau_{xy}^2\right]} \quad (3.12)$$

$$\tan 2\alpha = \frac{2\tau_{xy}}{\sigma_x - \sigma_y}.$$

The maximum shear stress acts on planes with axes at 45° to principal axes and is given by

$$\hat{\tau} = \frac{\sigma_I - \sigma_{II}}{2} = \sqrt{\left[\left(\frac{\sigma_x - \sigma_y}{2}\right)^2 + \tau_{xy}^2\right]}. \quad (3.13)$$

3.4 RIGOROUS ANALYSIS OF DISTRIBUTION OF FORCES

Manipulation of equations (3.1)–(3.9) for the analysis of the distribution of moments, etc. throughout a slab is complex. Rigorous solutions have been obtained for a few simple shapes of plate under particular load distributions [3–7], but no generally applicable method of rigorous solution has been found. Furthermore, no bridge deck rigorously satisfies the assumptions of isotropic or orthotropic behaviour with the result that assumptions of simplified structural action are necessary to interpret structural details into mathematical stiffnesses. Thus it is in general both impossible to develop rigorous mathematical equations to represent a structure and also impossible to solve the equations once obtained. However, approximate methods are available which either solve the plate bending equations by approximate numerical methods, or which represent the two-dimensional continuum of the deck by an assemblage of small elements or a grillage of beams. The latter methods, which have only been practical since the advent of the computer, have the advantages of direct physical significance to engineers and versatility in representing the different shapes, stiffnesses and support systems throughout a structure.

3.5 GRILLAGE ANALYSIS

Grillage analysis is probably the most popular computer-aided method for analysing bridge decks. This is because it is easy to comprehend and use, relatively inexpensive, and has been proved to be reliably accurate for a wide variety of bridge types. The method, pioneered for computer use by Lightfoot and Sawko [8] represents the deck by an equivalent grillage of beams as in Fig. 3.8. The dispersed bending and torsion stiffnesses in every region of the slab are assumed for purpose of analysis to be concentrated in the nearest equivalent grillage beam. The slab's longitudinal stiffnesses are concentrated in the longitudinal beams while the transverse stiffnesses are concentrated in the transverse beams. Ideally the beam stiffnesses should be such that when prototype slab and equivalent grillage are subjected to identical loads, the two structures should deflect identically and the moments, shear forces and torsions in any grillage beam should equal the resultants of the stresses on the cross-section of the part of the slab the beam represents. This ideal can in fact only be approximated to because of the different characteristics of the two types of structure summarized below.

Fig. 3.8 (a) Prototype deck and (b) equivalent grillage.

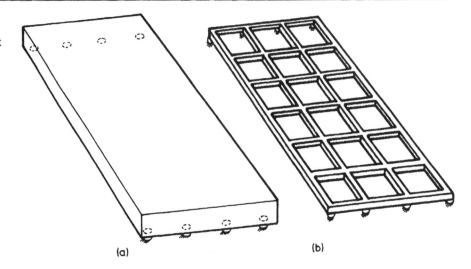

Firstly, equilibrium of any element of the slab requires that torques are identical in orthogonal directions, as equation (3.4), and also the twist $\partial^2 w/\partial x \partial y$ is the same in orthogonal directions. In the equivalent grillage there is no physical or mathematical principle that makes torques or twists automatically identical in orthogonal directions at a joint. However, if the grillage mesh is sufficiently fine, as in Fig. 3.9(a), the grillage deflects in a smooth surface with twists in orthogonal directions approximately equal (as will be the torques if torsion stiffnesses are the same in the two directions). On the other hand if the mesh is too coarse, as in (b), the grillage will not deflect in a smooth surface so that twists and torques are not necessarily similar in orthogonal directions. Even so it is often found that a coarse mesh is sufficient for design purposes.

Another shortcoming of the grillage is that the moment in any beam is solely proportional to the curvature in it, while in the prototype slab the moment in any direction depends, as in equation (3.5), on the

Fig. 3.9 (a) Fine and (b) coarse grillage meshes.

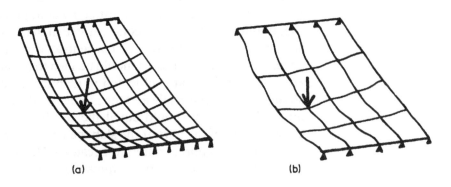

Grillage analysis

curvatures in that direction and the orthogonal direction. Fortunately it has been found from numerous comparisons of grillage with experiments and more rigorous methods (such as by West [9]) that bending stresses deduced from grillage results for distributed moments are sufficiently accurate for most design purposes. In the immediate neighbourhood of a load, which is concentrated in an area much smaller than the grillage mesh for the deck, the grillage cannot give the high local moments and torques, and influence charts or a local grillage is required as described in Sections 3.8 and 4.6.

3.5.1 Grillage mesh

Because of the enormous variety of deck shapes and support arrangements, it is difficult to make precise general rules for choosing a grillage mesh. However it is worth summarizing some of the deck and load characteristics that should be borne in mind.

1. Consider how the designer wants the structure to behave, and place grillage beams coincident with lines of designed strength (i.e. parallel to prestress or component beams, along edge beams, along lines of strength over bearings, etc.)
2. Consider how forces distribute within the prototype. For example if the deck has a cross-section as Fig. 3.10, torsion shear flows are as shown. The vertical shear flows at the edges of the slab are represented by components S_x of vertical shear forces in edge grillage members. For the prototype/grillage equivalence to be as precise as possible, each edge grillage member should be close to the resultant of the vertical shear flows at the edge of the deck. For a solid slab this is about 0.3 of the depth from the edge.
3. The total number of longitudinal members can be anything from one (if the slab is narrow enough to behave as a beam) to twenty or so (if

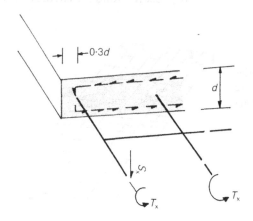

Fig. 3.10 Torsion forces at edge of grillage.

the deck is very wide and the design critical enough to warrant expense and trouble). There is little point in placing members closer than 2 to 3 times slab depth since local dispersion of load within the slab is not considered. On the other hand, if output information is to illustrate local high values, the maximum separation of longitudinal members for isotropic slabs should not be more than about ¼ of the effective span. For orthotropic slabs, the charts of Chapter 10 can be used to adopt a spacing so that a member with point load above carries not more than 40% of the load.

4. The spacing of transverse members should be sufficiently small for loads distributed along longitudinal members to be represented with reasonable accuracy by a number of point loads, i.e. spacing less than about ¼ of the effective span. In regions of sudden change such as over internal support, a closer spacing is necessary, as in Fig. 3.11. A separate local grillage can be used to study local effects, as in Fig. 3.11(b) and Fig. 4.11, with edge loads derived from the global grillage.

5. The transverse and longitudinal member spacings should be reasonably similar to permit sensible statical distribution of loads.

6. Simply supported decks at skew angles less than 20° can usually be analysed with grillages having right supports. However for a higher angle of skew, or if the deck is continuous, the lines of the grillage supports should be within about 5° of the skew supports of the prototype.

7. In general, transverse grillage members should be at right angles to longitudinal members (even for skew bridges) unless, as discussed in Chapter 9, directions of strength such as reinforcements are skew.

8. If the deck is at high skew or bearings are close together, the compressibility of the bearings has considerable effect on local shear forces, etc. and so should be represented with care.

9. It is implicitly assumed in a grillage analysis that point loads represent loads distributed over the width represented by the

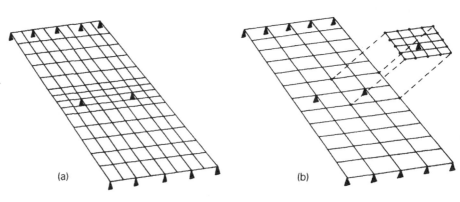

Fig. 3.11 Grillage meshes at internal supports: (a) local fine mesh and (b) separate course and fine grillages.

Fig. 3.12 Subdivision of slab deck cross-section for longitudinal grillage members.

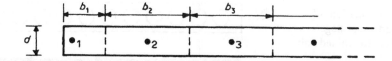

member. Sometimes decks with isolated point supports are best studied with two independent grillages. The first, with a coarse mesh of the whole deck, is used to study distribution of moments, etc. between spans; the second, with the finer mesh, represents only a small region around the support. The forces and displacements applied to the boundaries of this smaller grillage are derived from the forces and displacements output for the same points in the coarse grillage.

3.5.2 Grillage member section properties

Bending inertias

The bending inertias of longitudinal and transverse grillage members are calculated by considering each member as representing the deck width to midway to adjacent parallel members as shown in Fig. 3.12. The moment of inertia is calculated about the neutral axis of the deck. Thus for an isotropic slab,

$$I = \frac{bd^2}{12}. \qquad (3.14)$$

If the deck has thin cantilever or intermediate slab strips as Fig. 3.13, the longitudinal members can be placed as in (a) or as in (b). In (a) the inertias of all members are calculated about the deck neutral axis. However if the grillage members are placed as in (b), the thin slabs above members 1, 5 and 9 act primarily as flanges to members 2, 4, 6 and 8, respectively. Consequently the inertias of 1, 5 and 9 are calculated

Fig. 3.13 Alternative positions for longitudinal grillage members for deck with thin cantilever and connecting slabs.

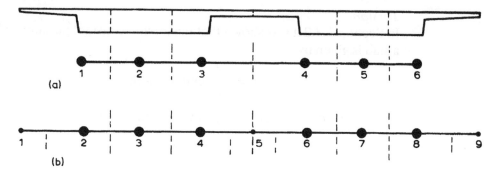

Fig. 3.14 Positions of longitudinal grillage members for voided slab decks.

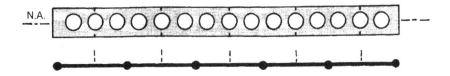

about centroid of the thin slab, while members 2, 4, 6 and 8 are calculated with the flanges as in (a) but with small inertias of 1, 5 and 9 deducted. Transversely, the thin slab flexes about its own centroid so that the thin slab depth is used in equation (3.1) for members 1–2, 4–5, 5–6, 8–9, while the thick slab depth is used for 2–3, 3–4, 6–7 and 7–8.

For a voided slab deck such as Fig. 3.14 the longitudinal grillage member inertias are calculated for shaded section about the neutral axis. Transversely, the inertia is generally calculated as at the centre line of void. However for void depths less than 60% of the overall depth, the transverse inertia can usually be assumed to be equal to the longitudinal inertia per unit width. Neither calculation is precise, but both are sufficient for design purposes.

If the moment–curvature equation (3.5) for a slab is compared with equation (2.3) for a beam, it can be seen that the slab equation differs not only due to the effect of transverse curvature but also because the effective stiffness is $1/(1-\nu^2)$ times that of the beam. This factor of increased stiffness of slab over equivalent beam is generally ignored in grillage analysis because both longitudinal and transverse stiffnesses are affected by the same relative amount and so do not alter distributions of load.

Reinforced and prestressed concrete slab bridges often have similar stiffnesses in longitudinal and transverse directions with the result that sufficient accuracy is obtained assuming that the full uncracked concrete section is effective, with reinforcing steel ignored. However if the transverse reinforcement is light while longitudinally the bridge is prestressed or heavily reinforced, account should be taken of flexural cracking, and the inertias in the two directions calculated separately for the different transformed sections.

Torsion
In Section 3.3.2 it was shown that the torsion constant per unit width of a slab is given by

$$c = \frac{d^3}{6} \text{ per unit width.}$$

Thus for a grillage beam representing width b of slab,

$$C = \frac{bd^3}{6}. \tag{3.15}$$

This is twice the magnitude of the moment of inertia given by equation (3.14), and in general it is possible to assume $C = 2I$ for grillage members representing slabs. There is no simple rigorous rule for calculating C for voided slabs and the above rule of $C = 2I$ is as convenient and accurate as any.

In true orthotropic slabs, the torques in transverse and longitudinal directions are equal by equation (3.4) and at the same time both twists are identically equal to $\partial^2 w/\partial x \partial y$. Consequently the transverse and longitudinal grillage members should have identical torsion constants per unit width of deck. Following the approximation of Huber in equation (3.9) it is suggested that the torsion constant of transverse and longitudinal grillage beams be

$$c = 2\sqrt{(i_x i_y)} \qquad (3.16)$$

where c = torsion constant per unit width of slab, i_x = longitudinal member inertia per unit width of slab and i_y = transverse member inertia per unit width of slab.

In beam-and-slab construction and 'orthotropic' steel battledeck construction, torques are not the same orthogonally and equation (3.16) does not apply (see Chapter 4 and [2]), and torsion constants are different.

Fig. 3.15 Torsion of (a) slab; (b) equivalent grillage and (c) forces in part of the grillage.

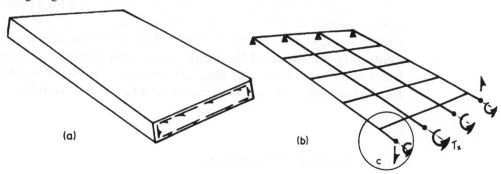

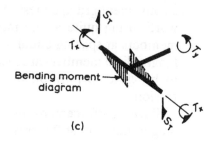

At the edges of a slab, the resultant horizontal shear flows near top and bottom faces (Fig. 3.10) terminate short of the edge of the slab at a distance of approximately 0.3 of depth. The equivalence of grillage and prototype is improved if the width of edge member is reduced for calculations of C to $(b - 0.3d)$.

It was mentioned in Section 3.3.2 that 'torque' in a slab describes only the torque on a section due to opposed horizontal shear flows near top and bottom faces, while the vertical shear flows at the edges are considered as part of vertical shear forces. The grillage reproduces the behaviour very closely. Figure 3.15 shows the slab of Fig. 3.5 in (a) together with equivalent grillage in (b). The forces on the cross-section in (b) are equivalent to those in (a) with grillage member torques T_x equivalent to torque in slab due to opposed horizontal shear flows while shear forces S_T are equivalent to vertical shear forces in slab. The reason the shear force S_T is generated in edge members of the grillage (and at edges of slab) by torsion is demonstrated in (c). Twisting of the grillage induces twisting and torques in both longitudinal and transverse members. At the joint between a transverse member and the edge longitudinal member, the transverse torque reacts with bending moments and shear forces S_T in the longitudinal member. At internal joints most of the transverse torque passes across the joint, and only the small difference in transverse torques on the two sides reacts with bending and shear longitudinally.

One exception to the above rule occurs if the grillage for a beam-like bridge is devised with only one structural longitudinal member and various outrigged transverse members. Since the longitudinal member must carry the whole of the torque on cross-section due to opposed horizontal shear flows and opposed vertical edge shear forces, the torsion constant must be calculated as for a beam in Section 2.4.

3.6 GRILLAGE EXAMPLES

3.6.1 Solid slab

Figure 3.16 shows a single-span solid reinforced concrete slab deck and a convenient grillage mesh. Since the grillage is small, as is its cost, it is worth using quite a fine mesh here with transverse and longitudinal members at spacing equal to approximately 2½ times the depth. Edge longitudinal members are located at a distance of 0.3 of depth from edge to be close to position in prototype of vertical shear forces due to torsion.

Since reinforcement in two directions is of the same order of magnitude, the stiffnesses will be assumed equal so that the slab is

Fig. 3.16 Grillage of solid slab deck: (a) section and (b) plan.

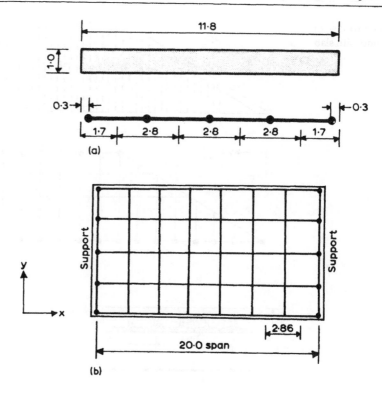

isotropic. By calculating the inertias on the full concrete area (i.e. ignoring cracking) we obtain

$$i_x = i_y = \frac{1.0^3}{12} = 0.0834 \text{ per unit width of slab in member}$$

$$c_x = c_y = \frac{1.0^3}{6} = 0.167 \text{ per unit width of slab in member.}$$

For internal longitudinal members with widths of 2.8 m we obtain

$$I_x = 2.8 \times 0.0834 = 0.233 \quad C_x = 2.8 \times 0.167 = 0.466.$$

For the edge longitudinal member the width is 1.7 m for calculation of I and $(1.7 - 0.3) = 1.4$ for calculation of C (i.e. width subjected to horizontal torsional shear flows.) Hence

$$I_x = 1.7 \times 0.0834 = 0.142 \quad C_x = 1.4 \times 0.167 = 0.233.$$

Transverse inertias are calculated in the same way.

3.6.2 Composite solid slab with edge stiffening
Figure 3.17 shows a single-span slab deck constructed compositely of precast prestressed concrete beams with in-fill reinforced concrete. The

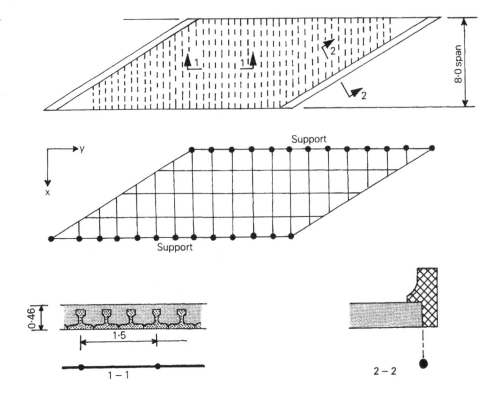

Fig. 3.17 Grillage of skew composite slab deck.

beams span right to the abutment and at the high-skew edges they are supported by an edge beam forming part of the parapet.

The grillage mesh has been chosen so that longitudinal members are parallel to prestressed beams with the transverse members at right angles. Each longitudinal member represents three prestressed beams, while transverse members are placed at ¼ span spacing. The edge member is concentric with the centre line of the edge stiffening upstand.

The in-fill concrete has lower strength and stiffness than the prestressed concrete so that it has a modular ratio $m = 0.8$ compared to prestressed concrete. Differing transformed cracked section inertias are used in the two directions because the transverse reinforcement is light. Furthermore, this reinforcement has different areas in top and bottom mats so that the transverse transformed section has different inertias for sagging and hogging; the average of the two is used. As a result with $m = 7$ for reinforcement (short-term loading),

$$i_x = 0.0070 \quad \text{per unit width of slab in member}$$
$$i_y = 0.00035 \quad \text{per unit width of slab in member.}$$

Applying equation (3.16),

$$c_x = c_y = 2\sqrt{(i_x i_y)} = 0.0031 \quad \text{per unit width of slab in member.}$$

The section properties of the edge stiffening beam are calculated for the area shown cross-hatched in Fig. 3.17. (It is assumed that the effect of the composite slab is included in internal orthogonal grillage members.) Since the edge stiffening behaves like a beam, the torsion constant is calculated as in Section 2.4. It should be noted that the inertia of the edge beam has not been increased here in the manner described in Section 8.3 because the transverse stiffness of the composite slab may not be sufficient to act as an effective flange to the upstand.

3.6.3 Two-span voided slab

Figure 3.18 shows a two-span voided slab deck with edge cantilever slabs supporting services. The grillage has four longitudinal members. The two-edge members are concentric with the centre line of edge webs in which torsion vertical shear flows will be concentrated. Internal longitudinal members pass through bearing positions. The transverse members are in general orthogonal to the longitudinal members and at approximately 1/5 effective span centres. Near the internal support they are closer to permit analysis of the sudden variations. In addition a skew member is placed between the bearings to represent the concentration of strength in the form of diaphragm beam reinforcement.

The longitudinal inertia is calculated for the deep section in Fig. 3.18

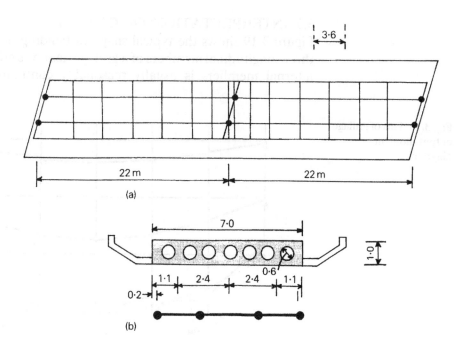

Fig. 3.18 Grillage of continuous voided slab deck: (a) plan and (b) section.

72 Slab decks

which gives

$$i_x = \frac{1.0^3}{12} - \frac{\pi \times 0.6^4}{64} = 0.077 \quad \text{per unit width of slab in member.}$$

Assuming the deck behaves isotropically,

$$i_y = i_x$$
$$c_x = c_y = 2i_x = 0.154 \quad \text{per unit width of slab in member.}$$

Thus we find for the internal longitudinal members

$$I_x = 2.4 \times 0.077 = 0.185 \quad C_x = 2.4 \times 0.154 = 0.37$$

while for the edge members

$$I_x = 1.1 \times 0.077 = 0.085 \quad C_x = 0.9 \times 0.154 = 0.14.$$

It will be seen that the torsion constant is calculated only for the width 'inside' the edge vertical shear forces (here coincident with the edge member). For orthogonal transverse edge members near midspan

$$I_y = 3.6 \times 0.077 = 0.277 \quad C_y = 3.6 \times 0.154 = 0.55.$$

Determination of the section properties of the skew diaphragms is imprecise. It is suggested that it should be based on the amount of reinforcement in the diaphragm in excess of that distributed as elsewhere in the slab.

3.7 INTERPRETATION OF OUTPUT

Figure 3.19 shows the typical shape of bending moment diagrams for three longitudinal members of a grillage near one edge. The diagram for internal members is usually reasonably continuous and the design

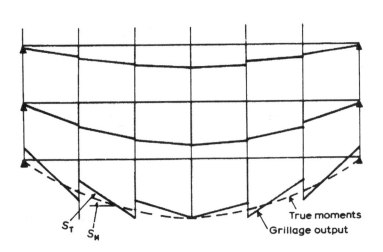

Fig. 3.19 Part of grillage output moment diagram.

moments can be read straight off the grillage output. The edge member diagram is typically discontinuous with 'saw teeth' because of the effects of torsion illustrated in Fig. 3.15. The saw tooth diagram of Fig. 3.19 can be thought of as the superposition of a saw tooth moment diagram due to torsion on a continuous moment diagram, shown dotted, due to bending. As a result, the bending moments should be taken as the average of grillage output moments on each side of a node.

The shear force output for each grillage member is the slope of the output saw tooth bending moment diagram. It includes both the component due to true bending S_M and the component due to torsion S_T as shown in Fig. 3.19. Since both components act together in the prototype slab it should be designed to support the full bending and torsional shear force as output by the grillage.

The torque in a true orthotropic slab is the same in orthogonal directions; however it is often different when read from grillage output. The design torque at any point should be taken as the average of those output for local transverse and longitudinal members per unit width of slab.

3.8 MOMENTS UNDER CONCENTRATED LOADS

The effective area of application of a concentrated load can be assumed to spread outwards through the surfacing and slab to the plane of the neutral axis as shown in Fig. 3.20. This dispersal is often assumed to occur at a spread-to-depth ratio of 1 horizontally to 2 vertically through the surfacing, and 1 horizontally to 1 vertically through the structure. If this area of application is equal to or larger than the grillage mesh (or if the areas due to several loads touch and together are larger), the load can be assumed to be sufficiently dispersed for the grillage to reproduce the distribution of moments throughout the slab. No further modification of moments, etc. is necessary. On the other hand, if the area of application of the load is small compared to the grillage mesh, no information will be obtained about the local high values under the load, though the grillage distributed moment field will simulate that in the

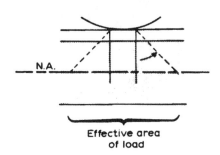

Fig. 3.20 Dispersion of concentrated load to plane of neutral axis.

Fig. 3.21 Shear-key deck.

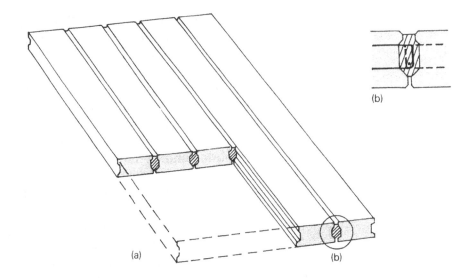

deck. The additional moments due to high local curvatures can be obtained for the area of slab within the grillage mesh from a local grillage or from influence charts such as those of Pucher [7].

3.9 SHEAR-KEY SLAB DECKS

A shear-key slab deck, shown in Fig. 3.21, is constructed of a number of parallel contiguous beams attached to each other along their length by stitch joints which have low transverse bending stiffness. The longitudinal joints can be thought of as full length 'piano' hinges. When the deck is loaded, as in Fig. 3.22, part of the load is carried by the beams

Fig. 3.22 Transverse load distribution by vertical shear resisted by beam torsion.

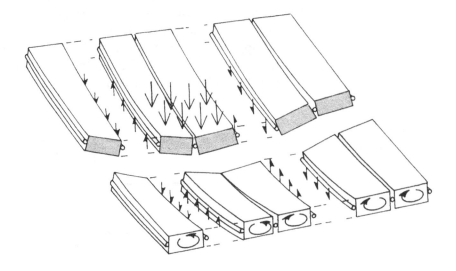

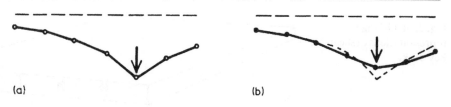

Fig. 3.23 Midspan deflections of shear-key deck supporting point or line load: (a) free joints and (b) partially stiff joints.

beneath and part transferred laterally to neighbouring beams by vertical shear forces on the hinges. Unlike a true slab deck this transverse shear is resisted primarily by the torsional stiffness of the beams and only to a small extent by the transverse bending stiffness of the articulated structure (which is nil if the joints are true hinges). If the joints really did behave as free hinges, the deck could be expected to deflect under a point load with a sharp cusp, shown in Fig. 3.23(a). The relative rotation of the beams on the two sides of the cusp is then so large that crushing damage of the road surfacing might be expected. In practice, the flexural stiffnesses of the joints are not negligible and the behaviour of the deck is significantly modified. Best [10] showed that with relatively little reinforcement in the shear keys, the relative rotation of the beams is considerably reduced below the levels predicted on the basis of flexible hinges, as illustrated in Fig. 3.23(b).

Shear-key slab decks are not common because the transverse load distribution is not nearly as effective as that of a slab. However they have been found useful for special situations, such as the replacement of a railway bridge, when the structure has to be erected and completed within a couple of days.

In practice it is difficult to predict the bending stiffness of some joint configurations and it is then sensible to assume for purposes of design that the joints are flexible. This results in conservative estimates for maximum beam torsions. None the less, unless the deck is somehow constructed with the real hinges, the ignored joint stiffness prevents damaging transverse crushing strains occurring in the surfacing.

Prior to the development of computer grillage methods shear-key decks were analysed with charts based on the 'articulated plate theory' of Spindel [11]. Grillage methods now provide a more versatile approach and are described below.

3.10 GRILLAGE ANALYSIS OF SHEAR-KEY SLAB

3.10.1 Grillage model

A shear-key deck such as in Fig. 3.24(a) can be represented by the grillage of (b) which has longitudinal members coincident with the centre lines of the beams of the prototype. Each longitudinal grillage

Fig. 3.24 Grillage representations of shear-key deck.

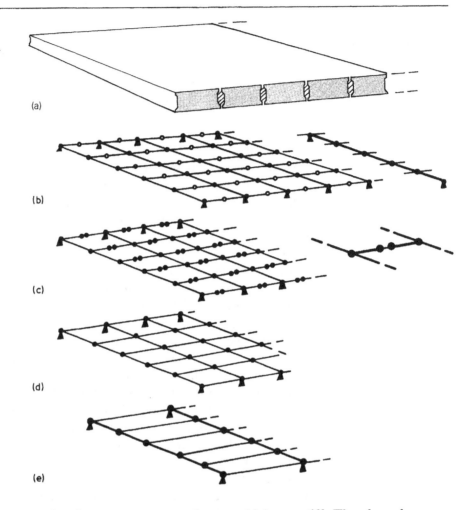

member has transverse outriggers which are stiff. The shear keys are represented by the pinned joints between the outriggers of adjacent beams.

If the shear keys do have bending stiffness (or if the computer program cannot accommodate pinned joints), the outriggers can be connected by short flexible members with all joints stiff as in (c). The length and stiffness of the flexible connection should simulate as accurately as possible the dimensions and stiffness of the shear key. For example, it is possible for the shear key to consist simply of steel dowels; in this case, the short flexible grillage members are given the effective length and stiffness of the dowels. One pitfall that can be encountered if a grillage has members with stiffnesses differing by several orders of magnitude is that the computer may not calculate with a sufficient number of significant figures so that solution of the stiffness equations may prove inaccurate or impossible.

The grillages of (b) and (c) have very large numbers of members with the result that the quantity of computer data and output is unmanageable. The number of members can be dramatically reduced in two ways. Firstly, the chains of transverse members of varying stiffness can be replaced by single members of equivalent stiffness as illustrated in (d) and explained below. Secondly, two or more deck beams can be represented by each longitudinal grillage member as in (e). The grillage members can be placed with lateral spacing of up to 1/5 of the span without seriously affecting the magnitude of calculated peak moments under a lane load. Such expediency might appear to reduce the accuracy of the analysis, but if it is done with care there is little discernible difference in the result, and the risk of arithmetic error is considerably reduced. Furthermore, when the overall analytical approximations are compared with the structure as built, it is often difficult to say which analysis is closest to the very different reality.

The chain of transverse members of various stiffnesses in Fig. 3.25(a) has the same stiffness under symmetric bending as the equivalent member of (b) if

$$\frac{M}{\theta} = \frac{2}{\int \frac{dx}{EI}} = \frac{2}{\frac{l-a}{E_1 I_1} + \frac{a}{E_2 I_2}} = \frac{2}{\frac{l}{E_e I_e}} \quad (3.17)$$

where $E_1 I_2$, $E_2 I_2$ are the flexural rigidities of the segments of the variable stiffness prototype, and $E_e I_e$ is the flexural rigidity of the equivalent grillage member.

Under the antisymmetric bending of (c) the equivalent member of (d) will have a different equivalent stiffness from that given by equation (3.17) unless shear flexibility of the equivalent member is considered.

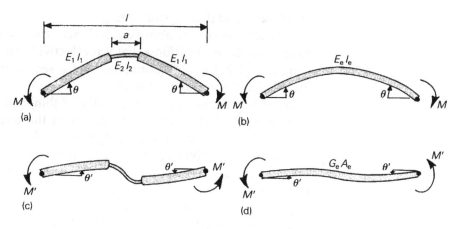

Fig. 3.25 Representation of chain of members with different inertias by equivalent uniform grillage member: (a) and (b) symmetric bending; (c) and (d) antisymmetric bending.

For the beams of varying inertia in (c) the stiffness is given by:

$$\frac{M'}{\theta'} = \frac{l^2}{2I_0} \qquad (3.18)$$

$I_0 = \int \frac{x^2 \, dx}{EI} =$ the second moment of area of dx/EI elements about their centroid.

$$I_0 = \frac{(l^3 - a^3)}{12 E_1 I_1} + \frac{a^3}{12 E_2 I_2}.$$

It should be noted that these equations assume that the beams of different inertia are symmetrically disposed about the midpoint.

For the equivalent shear flexible beam of (d) the stiffness is

$$\frac{M'}{\theta'} = \frac{l^2}{2\left(\dfrac{l^3}{12 E_e I_e} + \dfrac{l}{G_e A_e}\right)} \qquad (3.19)$$

where $A_e G_e$ is the shear rigidity of the equivalent grillage member.

Whence we obtain for equivalence

$$\frac{l^2}{2\left(\dfrac{l^3 - a^3}{12 E_1 I_1} + \dfrac{a^3}{12 E_2 I_2}\right)} = \frac{l^2}{2\left(\dfrac{l^3}{12 E_e I_e} + \dfrac{l}{G_e A_e}\right)}. \qquad (3.20)$$

Consequently, from equations (3.17) and (3.20) the section properties of the equivalent member are

$$\begin{aligned}
\frac{1}{I_e} &= \left(1 - \frac{a}{l}\right)\frac{E_e}{E_1 I_1} + \frac{a}{l}\frac{E_e}{E_2 I_2} \\
\frac{1}{A_e} &= \left[\left(1 - \frac{a^3}{l^3}\right)\frac{1}{E_1 I_1} + \frac{a^3}{l^3}\frac{1}{E_2 I_2} - \frac{1}{E_e I_e}\right]\frac{l^2 G_e}{12}.
\end{aligned} \qquad (3.21)$$

It will be found that A_e can be negative.

An exception to the above equivalence arises when the shear key has such a low flexural stiffness that it is effectively pin jointed. In this case equation (3.21) is indeterminate, and a real pinned joint should be included in the grillage member.

Shear-key decks can be constructed of thin-walled box beams whose cross-sections distort under the action of the transverse shear force. Such distortion can be included in the grillage analysis by suitably modifying the equivalent shear area of transverse members as described for cellular decks in Section 5.4.4. The simplest method of determining the equivalent shear area of the complicated section is to carry out a computer analysis of a plane frame of shape similar to the deck cross-section as demonstrated for a cellular deck in Fig. 5.15.

Fig. 3.26 Details for grillage of Fig. 3.24(e): (a) section of longitudinal member and (b) transverse member.

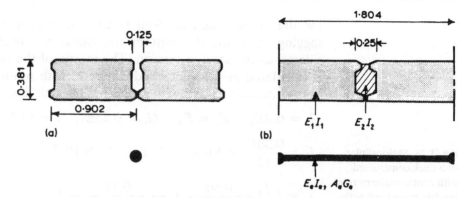

3.10.2 Example of section properties

Figure 3.26(a) shows the cross-section of the beams represented by each longitudinal grillage member in Fig. 3.24(e). The moment of inertia of the longitudinal member equals the sum of the inertias of the beams it represents. Here

$$I_x = 2 \times \frac{0.902 \times 0.381^3}{12} = 0.0083.$$

Since the joints do not provide continuity of torsion shear flow from one beam to the next, the torsion shear flow in each is a closed loop as in Fig. 3.27. This torsional behaviour is identical to that for an isolated beam for which the torsion constant is given in Section 2.4.3. The torsion constant of the grillage member is simply the sum of the St Venant torsion constants of the constituent beams. Here

$$C_x = 2 \times \frac{3 \times 0.902^3 \times 0.381^3}{10(0.902^2 + 0.381^2)} = 0.025.$$

It should be noted that this torsional behaviour is totally different from that of the orthotropic slab deck of Fig. 3.1(d) in which transverse prestress enables torsional shear flow interaction between beams. Then the torsion shear flow on the gross cross-section flows round the whole section as in Fig. 3.5.

The transverse grillage member must be equivalent to the varying transverse section of Fig. 3.26(b). The inertia of the joint is calculated

Fig. 3.27 Torsion shear flow in beams of shear-key deck.

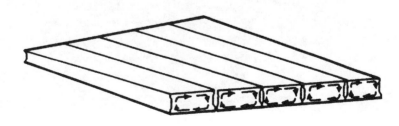

for the transformed cracked section of reinforced concrete under sagging moment, the bottom transverse reinforcement being considered as the steel in tension. The inertia of the beams on each side is calculated as for uncracked concrete, with reinforcement ignored. Using equation (3.21), with properties per unit length

$$E_2 = 0.8E_1, \quad E_e = E_1, \quad G_e = 0.43E_1, \quad l = 1.804 \quad a = 0.25$$

$$i_1 = \frac{0.381^3}{12} = 4.6 \times 10^{-3} \quad i_2 = 4.1 \times 10^{-4}$$

$$i_e = 1 \bigg/ \left(\frac{0.86}{4.6 \times 10^{-3}} + \frac{0.14}{0.8 \times 4.1 \times 10^{-4}} \right) = 1.65 \times 10^{-3}$$

$$a_e = 1 \bigg/ \left(\frac{0.997}{4.6 \times 10^{-3}} + \frac{0.0027}{0.8 \times 4.1 \times 10^{-4}} - \frac{1}{1.65 \times 10^{-3}} \right) \frac{1.804^2 \times 0.43}{12}$$

$$= -0.022$$

Fig. 3.28 Multicellular slab deck constructed with contiguous concrete top-hat beams and with steel piers acting compositely with deck, Westway, London; designed by G. Maunsell & Partners. Photograph courtesy G. Maunsell & Partners.

The joint has little transverse torsional stiffness and a nominal torsion constant can be attributed to transverse grillage members.

REFERENCES
1. Troitsky, M.S. (1967) *Orthotropic Bridges: Theory and Design*, The James F. Lincoln Arc Welding Foundation, Cleveland, Ohio.
2. American Institute of Steel Construction (1963) *Design Manual for Orthotropic Steel Plate Deck Bridges*, AISC, Chicago.
3. Timoshenko, S.P. and Woinowsky-Krieger, S. (1959) *Theory of Plates and Shells*, McGraw-Hill, New York, 2nd edn.
4. Rowe, R.E. (1962) *Concrete Bridge Design*, CR Books, London.
5. Jaeger, L.G. (1964) *Elementary Theory of Elastic Plates*, Pergamon Press, Oxford.
6. Bakht, B. and Jaeger, L.G. (1985) *Bridge Analysis Simplified*, McGraw Hill, New York.
7. Pucher, A. (1964) *Influence Surfaces of Elastic Plates*, Springer Verlag, Vienna and New York.
8. Lightfoot, E. and Sawko, F. (1959) Structural frame analysis by electronic computer: grid frameworks resolved by generalised slope deflection, *Engineering*, **187**, 18–20.
9. West, R. (1973) 'The use of a grillage analogy for the analysis of slab and pseudo-slab bridge decks', Research Report 21, Cement and Concrete Association, London.
10. Best, B.C. (1963) 'Tests of a prestressed concrete bridge incorporating transverse mild-steel shear connectors', Research Report 16, Cement and Concrete Association, London.
11. Spindel, J.E. (1961) 'A study of bridge slabs having no transverse flexural stiffness', PhD Thesis, King's College, London.

4 Beam-and-slab decks

4.1 INTRODUCTION

A large proportion of modern small- and medium-span bridges have beam-and-slab decks. This chapter describes how, for the purposes of design, they can be thought of as two-dimensional structures with behaviour which is in several ways simpler than that of slabs. They lend themselves to analysis by grillage with an equivalence of prototype and model which has immediate appeal to engineers. Examples are demonstrated for several types of structure. Occasionally for bridges which are not within the range of common practice it is necessary to

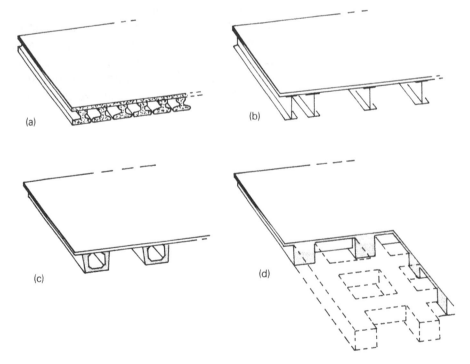

Fig. 4.1 Beam-and-slab decks: (a) contiguous; (b) spaced I-beams; (c) spaced box beams; and (d) grid.

investigate the secondary characteristics of such decks resulting from their truly three-dimensional form. These are mentioned at the end of the chapter and investigated further in Chapter 7.

4.2 TYPES OF STRUCTURE

The majority of beam-and-slab decks have a number of beams spanning longitudinally between abutments with a thin slab spanning transversely across the top as shown in Fig. 4.1. For short spans the beams are often contiguous as in (a), but for longer spans the beams are spaced as in (b) and (c). Transverse beams, called 'diaphragms', are placed to connect the longitudinal beams over the supports and sometimes at various sections along the span as in (d). The deck can have a high skew angle with beams displaced longitudinally relative to each other, and can be tapered with the beams not parallel. Curvature of the supported carriageway is usually accommodated by giving the edges of the slab the appropriate curvatures while supporting it on beams which are straight for each span. However, sometimes beams are also curved as described in Chapter 9.

4.3 STRUCTURAL ACTION

The behaviour of a beam-and-slab deck without midspan diaphragms as shown in Fig. 4.2 can be thought of as a simple combination of beams spanning longitudinally and slab spanning transversely. For longitudinal bending, the slab acts as top flange of the beams, and the deck can be thought of (and is sometimes constructed) as a number of T-beams connected along the edges of the flanges. Since the slab has a bending stiffness only a fraction of that of the beams, it flexes with much greater curvatures transversely than longitudinally, and, in spanning between beams, behaves much like a large number of transverse

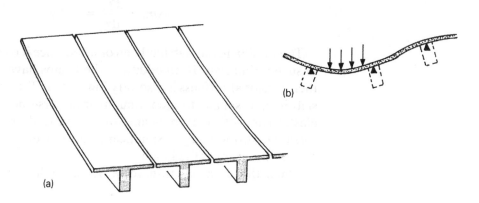

Fig. 4.2 Action of slab of beam-and-slab deck in (a) longitudinal bending as flanges of T-beams and (b) transverse bending as continuous beam.

Fig. 4.3 Element of beam-and-slab deck.

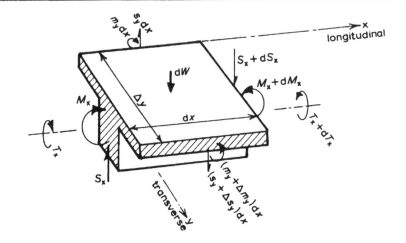

spanning planks. It is only in the immediate neighbourhood of a concentrated load that longitudinal moments and torques in the slab are of comparable magnitude to transverse moments. As shown in Section 4.6, it is generally possible to superpose the moments due to local two-dimensional dispersion of concentrated load on the transverse moments in the slab related to relative deflections and rotations of the beams.

Figure 4.3 shows an element of deck supporting an element dW of the locally dispersed load. The beam transmits moment M_x, shear force S_x and torsion T_x while the slab effectively only transmits transverse moment m_y and shear force s_y (per unit width of slab). The forces are related by equations

$$\frac{dS_x}{dx} + \Delta s_y = -W\Delta y$$

$$\frac{dM_x}{dx} = S_x \qquad (4.1)$$

$$\Delta m_y + \frac{dT_x}{dx} = s_y \Delta y.$$

The torsion in the slab has been omitted since it is comparatively small in such a thin slab. If, in addition, the beams have very thin I-sections their torsional stiffness is also very low so that T_x is effectively zero. The slab is then similar to a continuous transverse beam supported on an elastic support at each beam. In contrast, if the beams have high torsional stiffness, T_x is significant and the moments in the slab are discontinuous over the beam.

When the deck has transverse beams as in Fig. 4.4 the forces are

Fig. 4.4 Element of grid deck or of beam-and-slab deck at diaphragm.

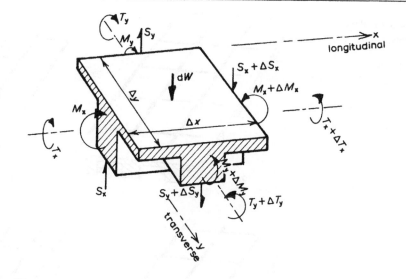

related by

$$\Delta S_x + \Delta S_y = -W\Delta x \Delta y$$
$$\Delta M_x + \Delta T_y = S_x \Delta x \tag{4.2}$$
$$\Delta M_y + \Delta T_x = S_y \Delta y.$$

These equations resemble those for a slab in Section 3.2. However the torques T in the two directions are not equal here, and depend on the differing twists and stiffnesses in the two directions.

In general, the stiffnesses of parts of the structure in longitudinal and transverse directions can be assumed the same as beams of similar cross-section, and stresses calculated from actions using Sections 2.3.2 and 2.4.3. Special considerations relating to slab behaviour and effective width of slab acting as flange to beams are discussed in the next section.

4.4 GRILLAGE ANALYSIS

4.4.1 Grillage mesh

Determination of a suitable grillage mesh for a beam-and-slab deck is, as for a slab deck, best approached from a consideration of the structural behaviour of the particular deck rather than from the application of a set of rules. Figure 4.5 shows four examples of suitable meshes for four types of deck.

In Fig. 4.5(a) the deck is virtually a grid of longitudinal and transverse beams. Since the average longitudinal and transverse bending stiffnesses are comparable, the distribution of load is somewhat similar

86 Beam-and-slab decks

Fig. 4.5 Grillage meshes.

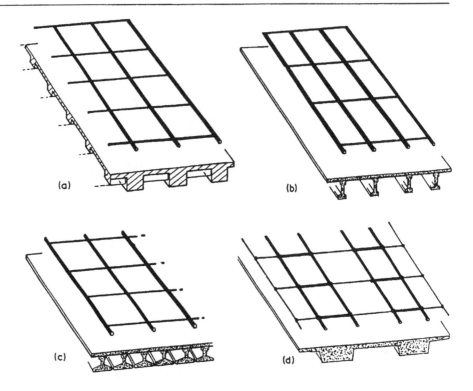

to that of a torsionally flexible slab, but with forces locally concentrated. The grillage simulates the prototype closely by having its members coincident with the centre lines of the prototypes beams.

The deck in Fig. 4.5(b) has longitudinal beams at centres a little less than the lane widths, and it is both convenient and physically reasonable to place longitudinal grillage members coincident with the centre lines of the prototype's beams. With no midspan transverse diaphragms the spacing of transverse grillage members is somewhat arbitrary, but about $\frac{1}{4}$ to $\frac{1}{8}$ of effective span is generally convenient. Where there is a diaphragm in the prototype such as over a support, then a grillage member should be coincident.

Figure 4.5(c) is a deck with contiguous beams at very close centres. Since a grillage with longitudinal members coincident with all beams can be unmanageable, it can be expedient to represent more than one beam by each longitudinal grillage member. However it is generally unwise to use a single grillage member to represent two beams of markedly different stiffnesses, because of the difficulty in apportioning the output. As computing facilities improve, it is becoming easier to use a fine mesh than to work out the characteristics of grillage members representing two or more beams. Since beam-and-slab decks have poor distribution characteristics, it is important not to place longitudinal grillage members much further apart than about 1/10 of the span, otherwise the concentration of moment will not be apparent in the grillage analysis. See also the spacing recommendations of Section 10.5.2.

The deck of Fig. 4.5(d) has large beams whose widths form a significant fraction of the distance between the centre lines. The longitudinal beams are then best represented as slabs, as described in Section 3.5.2, with two longitudinal members per beam. Since, during transverse flexure, the thin slab flexes much more than the thick beams, the grillage must also flex with most of the bending over the width of the thin slab. Accordingly the transverse members representing the thin slab will have much lower stiffnesses than the members representing beams. (An alternative approach is to use shear flexibility as described in Section 3.10 so that a single transverse member between longitudinal beam centre lines is able to represent the thick-thin-thick slab of the prototype. However, processing of the grillage output is then much more cumbersome.)

It is often convenient to place longitudinal members of nominal stiffness along the outer edges of the deck to define the overall width for loading.

4.4.2 Longitudinal grillage member section properties

Figure 4.6 shows part cross-sections of three beam-and-slab decks and the amount of each deck represented by the appropriate grillage member.

The flexural inertia of each grillage member is calculated about the centroid of the section it represents. Often the levels of the centroids of internal and edge member sections are at different levels. The significance of this is ignored unless a three-dimensional analysis is performed as described in Chapter 7.

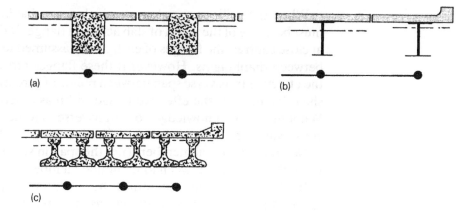

Fig. 4.6 Sections represented by longitudinal grillage members.

If the deck beams are spaced further apart than 1/6 of the effective span, or if the edge cantilever exceeds 1/12 of the effective span, shear lag significantly reduces the effective width of slab acting as flange to each beam. The grillage inertia should then be calculated using a reduced width of slab as described in Chapter 8.

Sometimes for the purpose of improving the simulation of applied loading demonstrated in Section 4.6, it is convenient to place longitudinal grillage members of nominal stiffness between those representing the structural sections of Fig. 4.6. The section properties of these members are calculated in a similar manner to that outlined in Section 3.5.2 for the deck in Fig. 3.13.

When the various decks in Fig. 4.6 are subjected to torsion, the 'beam' parts (dark shading in Fig. 4.6) behave like beams subjected solely to longitudinal torsion, while the 'slab' parts behave like slabs with torsion in both directions. Consequently the torsion constant C of the grillage member is the sum of the torsion constant of the beam calculated as in Section 2.4.3 and the torsion constant of the slab calculated as in equation (3.15). The torsion stiffnesses of some beams can be very low in comparison with the bending stiffnesses, as is the case for Fig. 4.6(b) and (c). It is then possible, for design purposes, to ignore torsion in the grillage, as is demonstrated in Section 4.9.

4.4.3 Transverse grillage member section properties

The section properties of a transverse grillage member, which solely represents slab, are calculated as for a slab. For this

$$I = \frac{bd^3}{12}$$

$$C = \frac{bd^3}{6}.$$

(4.3)

When the grillage member also includes a diaphragm, an estimate must be made of the width of slab acting as flange. If the diaphragms are at close centres, the flanges of each can be assumed to extend to midway between diaphragms. However if these flanges are wider than 1/12 of the effective transverse span between points of zero transverse moment, shear lag reduces the effective flange width as described in Chapter 8. Without prior knowledge of transverse moments, it is usually conservative to assume that the effective flange is 0.3 of the distance between longitudinal members (i.e. the span in shear lag calculations is twice the distance between longitudinal members).

If the construction materials have different properties in the longitudinal and transverse directions, care must be taken to estimate their relative stiffnesses. For example, a reinforced concrete slab on prestressed concrete or steel beams could be fully effective in compression for longitudinal sagging moments but behaves as a transformed cracked section for transverse bending. Furthermore it is possible that if the deck is continuous over a support, the same slab could be cracked for full depth in tension for longitudinal bending over the support so that only the reinforcing steel is effective. While an effort should be made to represent these differing characteristics, precise estimates of stiffness are seldom possible because of the unpredictable inelastic behaviour of construction materials. Information on torsion stiffness of cracked reinforced concrete is given in [1] and [2].

4.5 GRILLAGE EXAMPLES

4.5.1 Contiguous beam-and-slab deck

Figure 4.7 gives details of a single-span skew deck constructed of contiguous prestressed precast concrete beams with *in situ* reinforced

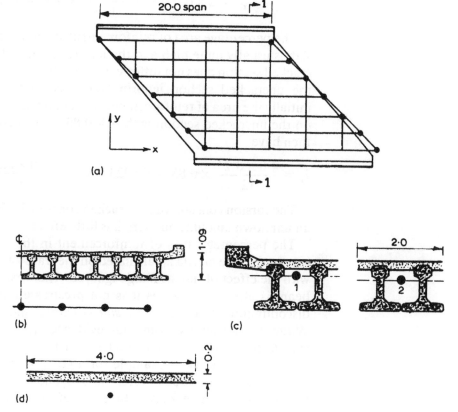

Fig. 4.7 Grillage of skew contiguous beam-and-slab deck: (a) plan; (b) part section 1–1; (c) longitudinal members; and (d) transverse member.

concrete slab. The grillage model has one longitudinal beam to each two prototype beams. While sufficient transverse members are provided for detailed analysis, their precise positions are chosen so that they intersect support beams at the same points as longitudinal beams. The skew of the grillage, but not the span, differs slightly from that of the prototype to make the mesh regular. (Such expediency greatly reduces the risk of human errors in calculations and thus is likely to improve the accuracy of analysis.)

The longitudinal members are calculated for sections in Fig. 4.7(c) assuming full concrete areas effective, but with *in situ* slab transformed by modular ratio $m = 0.85$ of Young's moduli for *in situ* and prestressed concrete.

$$I_{x1} = 0.24 \quad I_{x2} = 0.174$$

The torsion constants are calculated separately for the 'beam' and 'slab' parts of each member and added to give

$$C_{x1} = 2 \times 0.004 + \frac{2.0 \times 0.2^2 \times 0.85}{6} + 0.006 = 0.016$$

$$C_{x2} = 2 \times 0.004 + \frac{2.0 \times 0.2^3}{6} \times 0.85 = 0.010$$

The percentage area of reinforcement spanning transversely in such a slab is usually quite high and the inertia calculated on the transformed cracked section does not differ very much from the inertia calculated on the uncracked section ignoring reinforcement. Consequently, since initially the area of reinforcement is not known, the inertia is calculated on the uncracked section with $m = 0.85$. Transverse grillage members then have

$$I_y = \frac{4.0 \times 0.2^3}{12} \times 0.85 = 0.0023 \quad C_y = \frac{4.0 \times 0.2^3}{6} \times 0.85 = 0.0045.$$

The torsion constant of the cracked concrete is likely to be in error by an unknown amount, but here has little effect.

The percentage area of reinforcement in the support diaphragm is low, and so the inertia is calculated on the transformed cracked section, with the effect of slab as flange ignored. The torsion constant of such a beam is also very low, as it is not prestressed and it is made up of discontinuous sections of *in situ* concrete and precast beam web. Without relevant experimental evidence, it is suggested that C is calculated for the area of uncracked concrete used in calculating I.

A complete worked example for a composite prestressed concrete beam bridge is included in references [3] and [4]; and further advice on the design of such decks is included in references [5] and [6].

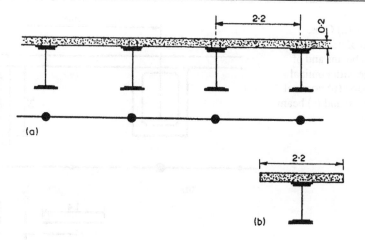

Fig. 4.8 (a) Cross-section of composite steel/concrete deck and of grillage and (b) section represented by longitudinal grillage member.

4.5.2 Spaced steel I-beams with reinforced concrete slab

Figure 4.8 shows part of a composite deck constructed of reinforced concrete slab on steel beams. Longitudinal grillage members are placed coincident with the centre lines of steel beams, and each represents the part of the deck section shown in (b). Using modular ratio $m = 7$ for steel (short-term loading) we obtain

$$I_x = 0.21$$

$$C_x = 0.000\,031 \times 7 + \frac{2.2 \times 0.2^3}{6} = 0.0032.$$

The slab is similar to that in Fig. 4.7 so that transverse grillage member properties are calculated in the same way.

Complete worked examples for composite steel beam bridges are included in references [7] and [8], and further advice on the design of such decks is included in references [9] and [10].

4.5.3 Spaced box beam with slab deck

Figure 4.9 shows the cross-section of a beam-and-slab deck constructed of spaced prestressed precast concrete box beams supporting a reinforced concrete slab. Longitudinal grillage members are placed coincident with centre lines of beams, with additional 'nominal' members running along centre lines of slab strips.

The section properties of the nominal members are calculated for width of slab to midway to neighbouring beams, hence

$$I_x = 1.4 \times \frac{0.25^3}{12} = 0.0018 \quad C_x = \frac{1.4 \times 0.25^3}{6} = 0.0036.$$

The properties of the beam members are calculated for the sections with flanges including the area in nominal members (unless shear lag has

Fig. 4.9 (a) Cross-section of deck with spaced beams and of grillage with nominal members; (b) nominal members; and (c) beam member.

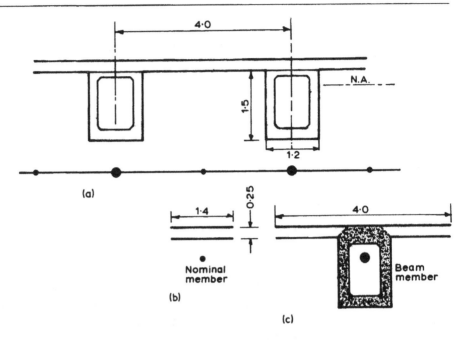

reduced the effective width of flanges to less), but with previously calculated properties of 'nominal' members deducted:

$$I_x = 0.57 - 2 \times \frac{0.0018}{2} = 0.57$$

$$C_x = 0.34 - 2 \times \frac{0.0036}{2} = 0.34.$$

Transverse members are calculated as in previous examples.

If the beams are much wider than those in Fig. 4.9 in comparison with the beam spacing, account must be taken of the variation in transverse flexural stiffness between slab and beam. If the walls are thin, distortion of the cross-section must be considered as discussed in Chapter 6. For decks with a few large cells it is simplest to use the techniques described in Chapter 5 for cellular decks using one longitudinal grillage member for each web and shear flexibility to simulate cell distortion. On the other hand, if the deck has a large number of beams it may be simplest to treat it as a shear-key deck as in Section 3.10, but using a plane frame analysis (see Fig. 5.15) to determine equivalent transverse grillage properties.

4.6 APPLICATION OF LOAD

The application of the loading and the interpretation of the output need to take account of the simultaneous global behaviour of the deck and the

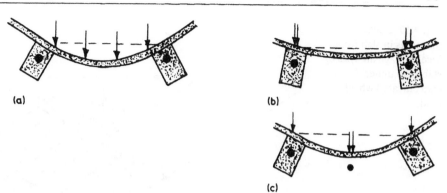

Fig. 4.10 Errors due to statical distributions of loading: (a) loading; (b) erroneous statical redistribution; (c) improved statical redistribution with nominal member.

local bending of the slab. It is generally easier to carry out separate calculations for the global and local effects, than to use a very detailed grillage model which includes local details. However, thought has to be given to how the loading can be separated into global and local effects, and to how the calculated forces can be superimposed in a consistent manner.

A load standing between beams, as in Fig. 4.10(a), can only be represented in the equations in the global grillage analysis by forces and couples at the joints. Some computer programs carry out a simple local statical distribution of load as is shown in (b). Unfortunately beam-and-slab decks can be very sensitive to such transverse movement of load, and the overall deck deformation and distribution of moments can be markedly different for load cases of (a) and (b). It is best to ensure that the transverse spacing of grillage members is less than about 3/4 of the vehicle or lane load width. This can be done by placing 'nominal' longitudinal members midway between widely spaced beam members as shown in (c) (see also Section 4.5.3). Statical distribution of loads transversely does not then reduce twisting forces on beams.

The local two-dimensional system of moments and torques in the thin slab under a concentrated load is not given by the global grillage and must be obtained from a separate local grillage or finite element analysis, or from the equations of Westergaard (quoted in [11]) or more simply from influence charts [5, 12]. Figure 4.11 illustrates a local grillage model for the slab between two beams supporting a wheel load. Symmetry is used to keep the model small. The slab strip is assumed to have a span equal to the clear distance between edges of beams plus the effective depth of the slab. The slab is assumed to be fixed along its edges. The effective area of the concentrated load can be assumed to spread down through the surfacing and slab to the midplane of the slab as explained in Section 3.8. A local grillage has the advantage that it is easy to model different stiffnesses in transverse and longitudinal directions. The local grillage can also be modified easily for use for other

Fig. 4.11 Local moments in slab from local grillage analysis of slab strip under concentrated wheel load.

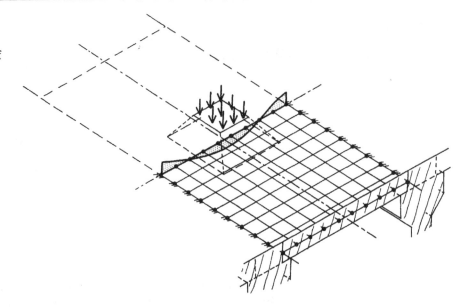

slab spans and details. The local moments derived from the local grillage must be added to those in the slab resulting from twisting and relative deflection of supporting beams. If there are no 'nominal' grillage members between beams and if transverse members have not been loaded, these moments can be read directly from grillage output for the local transverse member. If there is a 'nominal' longitudinal grillage member under the load or if transverse members have been loaded, the slab moments due to twisting of beams are best calculated from grillage output displacements and rotations of adjacent beams using equations (2.8).

4.7 INTERPRETATION OF OUTPUT

Figure 4.12 shows a typical bending moment diagram for part of a longitudinal beam of a beam-and-slab deck. Where transverse members represent only thin slab, the discontinuities in moment due to transverse

Fig. 4.12 Jump in grillage moment diagram for beam-and-slab deck at intersection of longitudinal beam and transverse diaphragm.

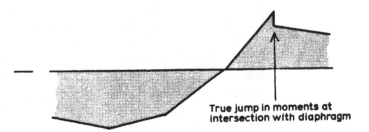

torques in the slab are small. The design moment should there be taken as the average of the moments on two sides of the joint (as for the slab deck in Section 3.7). In contrast, where the transverse member represents a diaphragm beam with significant torsion stiffness, the discontinuity in longitudinal moments is larger and represents a real change in moment across the joint. The design moments should then be taken as different on two sides of the joint and equal to grillage output moments. In a similar way, discontinuities in the moment diagram for transverse slab or diaphragm beams represent real changes in moment at connections with torsionally stiff longitudinal beams.

Design shear forces and torsions can be read directly from grillage output without modification.

Where the grillage member stiffness is calculated from properties of two distinct pieces of structure such as 'beam' and 'slab' in Fig. 4.6, the output torque (moment or shear) is attributed to each in proportion to its contribution to the particular stiffness. In general, stresses for prototype design should be calculated from the grillage output by applying the equations of Chapter 2 to the section assumed in the derivation of the grillage member properties.

4.8 TORSIONLESS DESIGN

The beams on many bridges have torsion stiffnesses which are low in comparison to bending stiffnesses. For example the concrete beams in Fig. 4.7 have torsion constants of less than 1/10 of the bending inertias, while the steel composite beams of Fig. 4.8 have torsion constants of less than 1/60 of the bending inertias. The design of these bridges can be simplified, in a safe manner, by ignoring the effects of torsion completely. When a deck is analysed with a 'torsionless' grillage the load distribution is not quite as effective as with a grillage with full torsion stiffnesses, and the calculated bending moments in the most heavily loaded beams are found to be slightly higher. If the beams are designed to carry the slightly higher moments from the torsionless design, they should be strong enough to carry the coexisting torques and moments calculated with a full torsion grillage.

Torsionless design should not be used when the beams have high torsion stiffnesses, such as the box beams in Fig. 4.9 which have torsion constants of the same order of magnitude as the bending inertias. The dividing line between bridges suitable and not suitable for torsionless design depends on the skew of the deck as well as the torsion stiffnesses and strengths of the beams. Each designer should check occasionally the relative magnitudes of the torsion strengths of beams and the torques from grillage output.

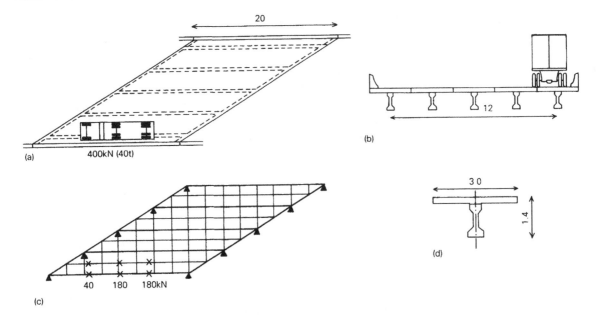

Fig. 4.13 Deck constructed with AASHTO type prestressed concrete girders: (a) plan; (b) section; (c) grillage; and (d) composite beam.

4.8.1 Spaced prestressed concrete beams

Figure 4.13 shows a composite bridge of 20 m span and 60° skew constructed of five prestressed concrete beams of AASHTO Type III at 3 m centres with 225 mm thick slab. The deck is analysed with the grillage with mesh as in Fig. 4.13(c), with each main longitudinal member representing the composite section shown in Fig. 4.13(d), while the nominal members in between are given the stiffness of a 1.5 m width of slab.

If the slab has a modular ratio of $m = 0.85$ relative to the beams, the longitudinal composite beams have section properties

$$\text{beams:} \quad I = 0.18 \, \text{m}^4.$$

The transverse members representing 2.5 m of slab have

$$\text{slab:} \quad I = 0.85 \times 2.5 \times 0.225^3/12 = 0.002 \, \text{m}^4.$$

The supports have vertical stiffnesses of 500 MN/m representing a rubber bearing on a stiff concrete structure.

Figure 4.14(a) illustrates the distribution of bending moments in the

Fig. 4.14 Distributions of moments from grillage of Fig. 4.13(c): (a) beams assumed torsionless and (b) full torsion.

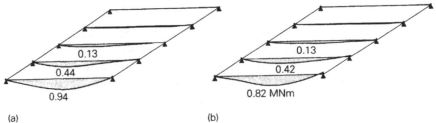

composite beams, ignoring torsion, calculated for a truck of 400 kN (40 tonne) in the position shown.

If a full torsion grillage is used the torsion constants of the members are

$$\text{beams:} \quad C = 0.018 \, \text{m}^4$$
$$\text{slab:} \quad C = 0.004 \, \text{m}^4$$

and the calculated bending moment diagram is as shown in Fig. 4.14(b). It is evident that the maximum moment from the full torsion grillage is about 13% lower than the moment from the torsionless grillage. (The maximum torque is about 6% of the maximum moment.)

If the skew is ignored and the structure analysed with a right grillage the maximum moment is found to be about 1.04 MNm from a torsionless grillage and about 0.93 MNm from a full torsion grillage (with maximum torsion of about 0.04 MNm). The relatively small difference between these various results illustrates how simply supported decks with widely spaced beams have relatively poor distribution characteristics, and are not very sensitive to skew, as is explained in Chapter 9. A hand calculation based on distribution coefficients for right decks, as explained in Chapter 10, might well be adequate in this case.

4.8.2 Spaced steel beams

Figure 4.15 illustrates a two-span continuous composite bridge with beams at 2.7 m centres, and skew of 20°. The deck is analysed with the grillage mesh of Fig. 4.15(c). The mesh has to be relatively fine in the region of the central support. In order to speed up data preparation a uniform mesh is used throughout. Also the transverse members are skewed, which is acceptable as explained in Section 9.1.2. Nominal longitudinal members have been placed between main members to simplify the application of loads.

Near the centres of the spans the composite beams are sagging and the section properties are calculated with the area of concrete in compression shown in Fig. 4.15(d). In the regions of the hogging over the internal supports (shown dotted in Fig. 4.15(c)) the section properties are calculated for the concrete cracked in tension and with the flanges reduced by shear lag, as explained in Chapter 8. The width of flange to each side of the shear studs in Fig. 4.15(e) is about 1/10 of the distance between points of zero moment. The flange stiffness in tension is calculated from the area of reinforcement within the effective flange width. The concrete is assumed to have $E = 30\,000$ MPa while the steel has $E = 200\,000$ MPa, so that the modular ratio of concrete relative to steel is 0.15. Using universal beams of $838 \times 292 \times 226$ kg and concrete slab 225 mm thick, the bending inertias of the composite beams in the

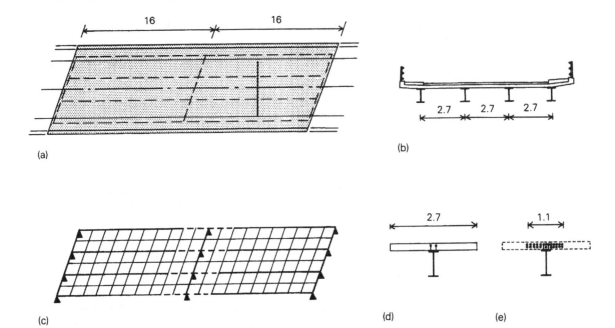

Fig. 4.15 Continuous composite deck with spaced steel beams: (a) plan; (b) section; (c) grillage; (d) composite beams for sagging; and (e) composite beam for hogging.

centres of the spans are (in terms of steel dimensions):

beams sagging: $I = 0.010 \, \text{m}^4$.

Over the supports slab stiffness is only provided by steel reinforcement within the effective flange width, so that the bending inertia of the composite beams is

beams hogging: $I = 0.004 \, \text{m}^4$.

The bending inertia of the slab can be calculated for the full concrete section, or for the 'net transformed' section with steel reinforcement in tension and concrete in compression. It is generally easier in design calculations to use the full concrete section. Here the bending inertia per unit width is

slab: $i = 1.0 \times 0.225^3/12 = 0.000\,95 \, \text{m}^4/\text{m}$.

If the steel Young's modulus is used for all members in the grillage, the concrete properties must be factored down by $m = 6.7$ to

transverse: $i = 0.000\,95 \times 0.15 = 0.000\,14 \, \text{m}^4/\text{m}$.

The nominal longitudinal members between composite beams are calculated with the same stiffness. The supports have vertical stiffnesses of 1000 MN/m.

Figure 4.16(a) illustrates the bending moment diagram calculated for distributed traffic loading and footpath load on both lanes of both spans, with knife-edge loads on the right span. It can be seen that the points of

Fig. 4.16 Distributions of moments from grillage of Fig. 4.15(c): (a) loading over whole deck and (b) loading on one side of one span.

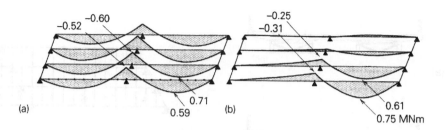

zero moment are at about 15–20% along the spans from the central supports. Figure 4.16(b) shows the moment diagram due to loading of the same intensity only on the bottom right of the deck in the diagram.

In a full torsion grillage the torsion constants of the grillage members are provided mainly by the concrete slab, and are approximately

$$\text{beams sagging:} \quad C = 0.0008 \text{ m}^4$$
$$\text{beams hogging:} \quad C = 0.0003 \text{ m}^4$$
$$\text{transverse:} \quad c = 0.000\,28 \text{ m}^4/\text{m}.$$

It is found that the maximum moments from the full torsion grillage are about 6% less than shown in Fig. 4.16(b). The coexisting torques in the steel beams are very low and do not require consideration in design. For this reason it is generally sensible to include torsion stiffness in grillages for composite steel decks.

If shear lag and cracking of the concrete slab are ignored over the central support, the section of Fig. 4.15(d) is used throughout the length of the longitudinal beams. It is then found that the support moment with both spans loaded is increased by more than 30%, while the sagging moment with one side of one span loaded is reduced by about 15%. The use of the reduced section in the initial analysis has had the effect of redistributing about 30% of the support moments.

The design of composite steel bridges is discussed in detail in references [7–10]. Complete worked examples are included in references [7] and [8].

4.9 BRACING

The composite steel beam deck in Section 4.8.2 has been analysed without consideration of transverse bracing. Bracing is commonly placed between pairs of beams as shown in Fig. 4.17. The bracing assists construction and restrains the bottom flanges where they are in compression near intermediate supports, as explained in references [7–10]. Some designers ignore discontinuous bracing in the grillage analysis for simplicity and on the assumption that to do so is conservative.

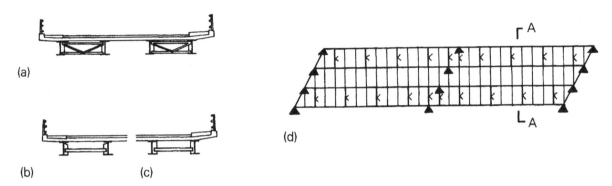

Fig. 4.17 Bracing of steel beams in pairs: (a) K-bracing; (b) and (c) transverse-beam bracing; (d) grillage model.

Others include the bracing in the grillage in order to check its loading and fatigue.

Discontinuous bracing presents a challenge for modelling by grillage because the grillage joints take no account of the transfer of forces from one type of member to another. Figure 4.18 illustrates the deformations of single frames of K-bracing and transverse-beam bracing under conditions of symmetric and antisymmetric bending. The moments on each frame from the adjoining structure come primarily from transverse bending of the slab and secondly from torsion, which is also mainly in the slab. If plane frame analyses are carried out of single frames of bracing and are subject to forces from the slab, as in Fig. 4.18, the bracing provides relatively little additional stiffness to the slab. In the case of Fig. 4.18(c) the bracing has no effect on the stiffness of the slab because the web stiffeners are purposely not attached to the top flanges. Bracing primarily restrains the lateral deflections of the beam bottom flanges: in so doing it interacts with the small warping torsional stiffnesses of the beams which are generally ignored in grillage analyses. If torsional warping is considered significant the analysis is best done with a space frame as explained in Chapter 7.

The results from grillages without and with bracing are compared in

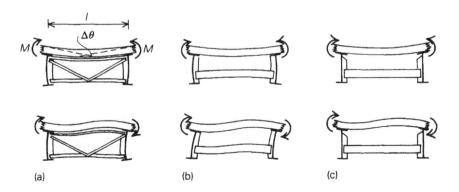

Fig. 4.18 Deformation of bracing under action of symmetric and antisymmetric components of transverse bending moments.

Fig. 4.19 Deflections and transverse moments across section A–A of grillage of Fig. 4.17(d) without and with bracing.

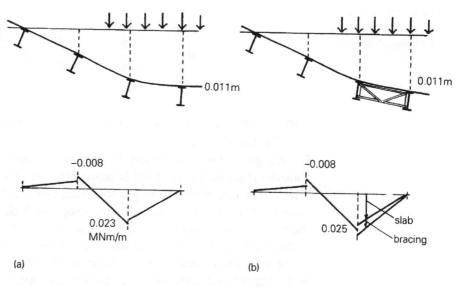

Figs. 4.19(a) and (b). Both diagrams show the deflections and transverse moments across section A–A of the grillage in Fig. 4.17(d) when loaded on one side. (Torsion stiffnesses were included.) The grillage with bracing had additional members to simulate K-bracing at each transverse member marked K. The stiffnesses were derived from plane frame models loaded as in Fig. 4.18(a). The plane frame included a narrow width of slab with width limited by shear lag to about 1/5 of the beam spacing. The moment of inertia I for the equivalent transverse member was calculated from the applied moments M and output relative rotation $\Delta\theta$ by

$$I = \frac{Ml}{E\Delta\theta}. \qquad (4.4)$$

It is evident in Fig. 4.19 that the discontinuous bracing has had relatively little influence on the transverse bending of the slab. The grillage output moments for the bracing are converted into bracing forces by back analysis with the plane frame.

When the bracing forms a continuous structure between several beams it can have a marked influence on load distribution and needs to be simulated in the grillage. It then interacts with vertical loads in the beams in addition to transverse flexure of the slab. The section properties of the grillage members representing K-bracing can be derived as transverse beams. However transverse-beam bracing may require much more detailed attention if it is flexible in shear, as shown in Fig. 4.20, and may warrant a shear-flexible grillage, as explained in Chapter 5 and Section 6.7, or a space frame analysis as in Chapter 7.

Fig. 4.20 Distortion of transverse-beam bracing under shear loading.

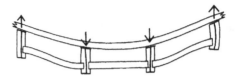

4.10 SLAB MEMBRANE ACTION IN BEAM-AND-SLAB DECKS

In the preceding discussion it has been tacitly assumed that for consideration of longitudinal bending, the slab can be thought of as a series of strips, each forming a top flange of a T-beam. No check has been made that after notionally cutting up the deck the displacements of the parts are compatible, i.e. that the parts can in fact be joined together without additional forces and distortion hitherto not considered.

Figure 4.21(a) shows the midspan section of a beam-and-slab deck with exaggerated deflections due to non-uniform loading. (b) shows the composite beams notionally separated, but with twists and deflections of (a). The grillage can adequately simulate these deflections and accompanying transfer of load by vertical shear and transverse bending of the slab. But inspection of the ends of the separate beams in plan or elevation, as in (c) and (d), shows that if all the beams flex about a neutral axis passing through their centroids, the ends of the slab flanges are displaced relative to each other. In reality this step displacement cannot happen, and the relative movement of the tops of the beams is resisted and reduced by longitudinal shear forces in the connecting slab

Fig. 4.21 Longitudinal 'warping' movement of slab of beam-and-slab deck: (a) and (b) span section; (c) support elevation; and (d) support plan.

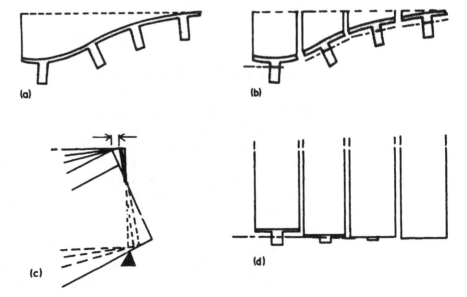

Fig. 4.22 Effects of slab membrane action in beam-and-slab deck: (a) in-plane shear in slab; (b) axial force in beam; and (c) movement of neutral axis.

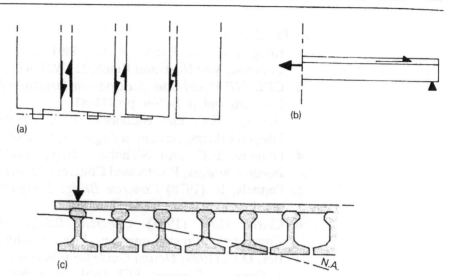

as shown in Fig. 4.22(a). These shear forces are in equilibrium with axial tension/compression forces in beams near midspan as shown in (b).

This transfer of shear force between beams with balancing axial forces cannot be simulated in a conventional grillage analysis. The forces have three effects on deck behaviour.

1. The shear forces in the slab can be much larger than predicted from grillage analysis.
2. The axial tension forces in the beams with largest deflections (i.e. under the load) cause the neutral axis to rise locally while compression forces elsewhere cause the neutral axis to move down, as shown in Fig. 4.22(c).
3. The load distribution characteristics of the deck are improved. The longitudinal interbeam shear forces and axial forces in Fig. 4.22 are at different levels and thus form couples which reduce the moment in the loaded beams and increase moments elsewhere.

It is often assumed that if a deck slab is subjected to shear forces in excess of its strength, designed from grillage analysis, it will only crack or yield and so relax the forces of Fig. 4.22. This may happen, but if the deck is outside the range of common practice, a three-dimensional analysis may be necessary as described in Chapters 7, 12 or 13.

The movement of the neutral axis in Fig. 4.22 occurs as a result of the difference in levels of the centroids of the beams and connecting slabs. The behaviour is further complicated if there are transverse beams also with centroids out of the plane of the slab. However, in contrast, the behaviour of cellular decks described in the next chapter is simpler because the centroid does not vary significantly in level between webs and slabs.

REFERENCES

1. Kong, F.K. and Evans, R.H. (1989) *Reinforced and Prestressed Concrete*, Van Nostrand Reinhold, Wokingham, 3rd edn.
2. *CEB-FIP Model Code for Concrete Structures* (1978) Comité Euro-International du Béton, pp 474–81.
3. Hambly, E.C. and Nicholson, B.A. (1990) Prestressed beam integral bridges, *Structural Engineer*, **68**, December.
4. Hambly, E.C. and Nicholson, B.A. (1991) *Prestressed Beam Integral Bridges*, Prestressed Concrete Association, Leicester.
5. Pennels, E. (1978) *Concrete Bridge Designer's Manual*, Cement and Concrete Association, London.
6. Clark, L.A. (1983) *Concrete Bridge Design to BS5400*, Construction Press, Longmans, London, with supplement (1985).
7. Iles, D.C. (1989) *Design Guide for Continuous Composite Bridges: 1 Compact Sections*, SCI Publication 065, Steel Construction Institute, Ascot.
8. Iles, D.C. (1989) *Design Guide for Continuous Composite Bridges: 2 Noncompact Sections*, SCI Publication 066, Steel Construction Institute, Ascot.
9. Johnson, R.P. and Buckley, R.J. (1986) *Composite Structures of Steel and Concrete*, Vol. 2 Bridges, Collins, London.
10. Hayward, A.C.G. (1988) *Composite Steel Highway Bridges*, British Steel General Steels, Motherwell.
11. Rowe, R.E. (1962) *Concrete Bridge Design*, CR Books, London.
12. Pucher, A. (1964) *Influence Surfaces of Elastic Plates*, Springer-Verlag, Vienna and New York.

Fig. 4.23 Steel plate girder composite bridge deck of Festival Park Flyover, Stoke-on-Trent, England. Deck designed by Cass Hayward & Partners, subconsultants to WS Atkins Consultants Ltd. Photograph courtesy of Bank House Studio.

5 Multicellular decks

5.1 INTRODUCTION

This chapter describes the modes of deformation and systems of internal forces that characterize the behaviour of multicellular decks. If the cells are stiffened with diaphragms or cross-bracing at frequent intervals, to prevent them changing shape by distortion, the deck can be analysed like a beam if it is narrow, or like a slab if it is wide. However it is often convenient and acceptable to use cellular structures and box-girders which do distort under shear and torsional loading, and it is then necessary to take account of the distortion in the methods of calculation.

The following sections describe the behaviour of wide cellular decks which can be analysed with a shear-flexible grillage. The shear-flexibility in the grillage reproduces the distortion behaviour of the cells. The application of this method, which was pioneered by Sawko [1], is demonstrated for a variety of structural forms. While the method is not the most theoretically rigorous analogy of cellular behaviour available, it has the benefit of being applicable to a wide variety of structures, of being relatively inexpensive in computer time and user time, and being relatively simple to comprehend [2]. The results of some analyses are presented and shown to compare well with more rigorous methods.

5.2 THE SHEAR-FLEXIBLE GRILLAGE

Figure 5.1 shows a number of cellular structures for which the shear-flexible grillage can be used. The analysis is most appropriate for wide multicellular decks with thin slabs enclosing rectangular cells as shown in Fig. 5.1(a). However, it has also been found to perform with acceptable accuracy for the analysis of some narrow decks and decks with inclined webs as in Fig. 5.1(c). It can also be used for decks with large cylindrical voids such as Fig. 5.1(d). Variations in structural depth or plate thicknesses can be included. The decks can also be curved or tapered in plan, but the analysis has the limitations described in Chapter

Fig. 5.1 Cellular decks.

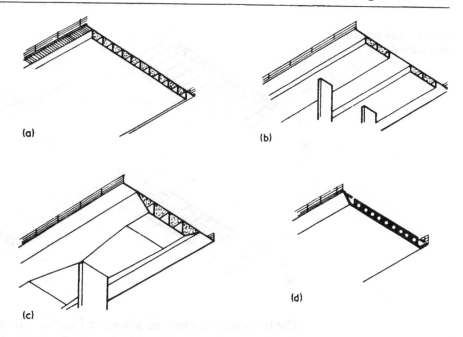

9. Transverse diaphragms can be placed anywhere at right angles or skew to the longitudinal webs.

5.3 GRILLAGE MESH

The following discussion of cellular behaviour and grillage simulation assumes that the grillage mesh is in the plane of the principal axis of bending of the deck as a whole with the longitudinal members coincident with longitudinal webs. Such positioning of longitudinal members is demonstrated in Fig. 5.2 for the decks of Fig. 5.1. This arrangement is chosen so that the web shear forces can be directly represented by grillage shear forces at the same points on the cross-section. Other arrangements are possible, but then the method of apportioning forces and stiffnesses to members is different from that described later. If the deck has sloping webs, the grillage simulation is not so precise and engineering judgement must be used to position longitudinal members. The members located along the edges of the side cantilevers are not generally necessary for the analysis of the cellular structure, but if they are included with nominal stiffnesses they can simplify the description of cantilever loading in the computer input. A single cell box can be analysed with a grillage having one longitudinal member to each web. With a shear-flexible grillage it is not appropriate to place nominal longitudinal members between webs, as is explained in Section 5.4.4.

108 Multicellular decks

Fig. 5.2 Grillage meshes for cellular decks.

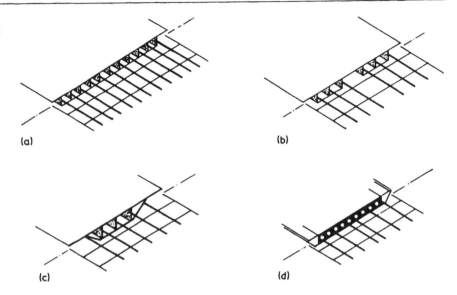

The transverse members, shown in Fig. 5.2, are spaced closer than ¼ of the distance between points of contraflexure and wherever there are diaphragms, unless they are too numerous. Wider spacing results in inaccuracy due to excessively large discontinuities in moments, etc. at the joints. Closer spacing results in more continuous structural behaviour and provides greater detail of forces, etc., but it does not make the characteristic behaviour of the grillage any closer to that of the cellular structure.

The underlying principle on which grillage member properties are derived is that when the grillage joints are subjected to the same deflections and rotations as coincident points in the structure, the member stiffnesses generate forces which are locally statically equivalent to the forces in the structure.

5.4 MODES OF STRUCTURAL ACTION

Figure 5.3 shows the displacement and deformation of a cross-section of a cellular deck under load split up into the four principle modes: longitudinal bending, transverse bending, torsion and distortion. The character and simulation of each of these modes is described below.

5.4.1 Longitudinal bending

Longitudinal bending behaviour can be visualized by notionally cutting the deck longitudinally between webs into a number of 'I-beams' as in Fig. 5.4. The longitudinal bending stresses on the cross-section in Fig. 5.5(a) are similar to those for the 'I-beam' subjected to the same

Fig. 5.3 Principal modes of deformation: (a) total; (b) longitudinal bending; (c) transverse bending; (d) torsion; and (e) distortion.

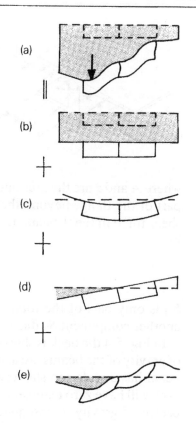

curvature as the deck, and are given by

$$\frac{\sigma}{z} = \frac{M}{I} = \frac{E}{R}. \quad (5.1)$$

The shear stress distribution due to bending is also similar to that from simple beam theory of 'I-beams' with transverse or longitudinal shear flow at the point given by

$$r = \frac{S_M A \bar{z}}{I} \quad (5.2)$$

Fig. 5.4 Cellular deck split up into I-beams by cuts along centres of cells.

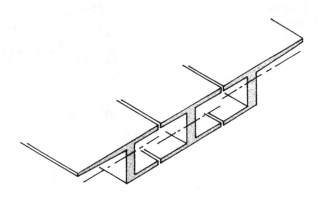

Fig. 5.5 Flexure of edge beam about its own neutral axis: (a) bending stresses and (b) shear flow.

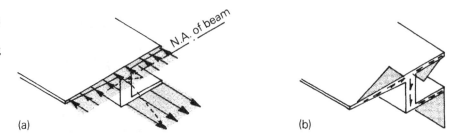

where A and $\bar{z}$ are the area and eccentricity of the centre of gravity of the part of the flange beyond the point of consideration. S_M is the vertical shear force in the 'I-beam' (i.e. in the web) due to bending and is given by

$$S_M = \frac{dM}{dx}. \qquad (5.3)$$

S_M is only part of the total vertical shear force in the web which has another component S_T due to torsion.

In Fig. 5.4 the deck is shown cut midway between webs. If the centres of gravity of the beams are at different levels as is the case when the edge beam has a wide top slab cantilever then, on being flexed separately, each will have zero extension along its own neutral axis at the level of its centre of gravity. Consequently, as shown in Fig. 5.6, cross-sections near the end would rotate about points at different levels and have a relative longitudinal displacement u. This displacement is resisted and reduced by the very high shear stiffnesses of the top and bottom slabs, and all the beams are forced to flex about a neutral axis which is virtually coincident with the principal axis of the deck as a whole. Consequently the section properties of each I-beam represented by a grillage member should be calculated about the principal axis of the deck as a whole. The

Fig. 5.6 Relative end displacement of beams with different neutral axes.

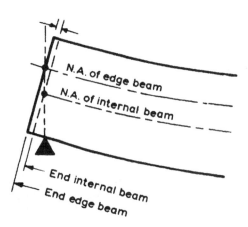

Fig. 5.7 Cellular deck split up into I-beams with common neutral axis.

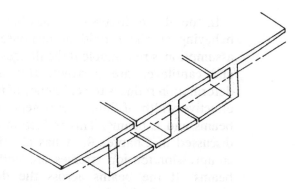

bending stresses are then calculated from these properties, using equation (5.1), with z measured from the common neutral axis.

Although the subdivision of the cellular deck into 'I-beams' is somewhat arbitrary, the stresses calculated as above from equation (5.1) can be considered accurate for practical purposes. A small change of the arbitrary position of the cuts changes the relative magnitudes of the grillage member inertias which in turn attract a different distribution of moments. But the ratio of M/I for any grillage beam remains effectively constant and the calculated bending stresses are not affected. However, shear flows calculated from equation (5.2) can only be considered approximate since the flange area A is arbitrary. These errors can be avoided if the deck is cut into 'I-beams' whose individual centres of gravity are on the principal axis of the deck. For the example of Fig. 5.7 this is done simply by cutting the slabs so that each web has identical proportions of top and bottom slab. On being flexed separately, the neutral axis of all beams will be on the same level and the distributions of bending stress and shear flow will be as in Fig. 5.8. As before, the bending stresses are described by equation (5.1) where I is the moment of inertia of the asymmetrical beam section of Fig. 5.7 and z is the distance from the neutral axis of the beam (and deck). Shear flows due to bending are now accurately described by equation (5.2) where A, $\bar{z}$ and I relate to the section of Fig. 5.7.

Fig. 5.8 Flexure of edge beams with centroid on neutral axis of deck: (a) bending stresses and (b) shear flow.

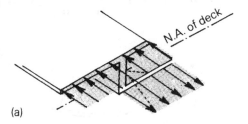

(a)

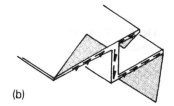

(b)

In the above discussion it has been assumed that the I-beams are behaving as shear rigid in response to longitudinal bending. This assumption is reasonable if the flanges are narrow (i.e. if the cell widths and cantilever are narrow). But if the flanges are wide, shear deformation reduces the efficiency of the sections which have a reduced effective width of flange, as is generally accepted in the design of T-beams and L-beams. This reduction of efficiency due to shear lag is discussed in Chapter 8. It has also been assumed that the bending compression/tension stresses are uniform across the width of the I-beams. If the beams across the deck are subjected to different curvatures, the bending stresses vary across the deck. This variation is continuous and, ignoring shear lag, can be assumed to be linear between webs. The stresses of equation (5.1) then represent the average stress across the 'I-beam'.

From the above discussion it is evident that the moments of inertia apportioned to longitudinal grillage members should be calculated about the principal axis of the deck as a whole. If possible, the grillage member should be made to represent part of the deck as in Fig. 5.7 which has its centre of gravity on this axis. However, for a deck such as Fig. 5.1(c), the top slab is very much wider than the bottom slab and such division of the cross-section gives beams of ridiculous cross-sections. In this case it is convenient to notionally cut the deck midway between the positions of longitudinal grillage members and accept some loss of accuracy. The moments of inertia and section moduli of members are still calculated about the principal axis of the deck as a whole.

5.4.2 Transverse bending

Transverse bending, shown in Fig. 5.9, is the flexure of top and bottom slabs in unison about a neutral axis at the level of their common centre of gravity, as if they were connected by a shear rigid web. It does not include the independent flexure of top and bottom slabs which results in cell distortion as shown in Fig. 5.3(e).

The moment of inertia of transverse grillage members is calculated about the common centre of gravity of the slabs and is

$$i_t = (h'^2 d' + h''^2 d'') = \frac{h^2 d' d''}{(d' + d'')} \quad \text{per unit length} \quad (5.4)$$

Fig. 5.9 Transverse bending.

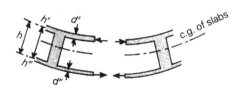

where d', d'', h', h'' are the slab thicknesses and distances from their centroid shown in Fig. 5.9. If the transverse grillage member also includes a diaphragm, the inertia should be calculated including the diaphragm.

Grillage analysis ignores the effects of Poisson's ratio on the interaction of longitudinal and transverse moments. In narrow decks this introduces little error as transverse moments are small and the deck is free to hog transversely, just like a beam under a longitudinal sagging moment. However in a wide deck with little stiffness against cell distortion, the calculated transverse moments can be considerably in error if they are small and if Poisson's ratio is significant. The transverse moment due to interaction of the longitudinal moment on the transverse moment can be large in comparison with the transverse moment due to transverse sagging. Since concrete has a relatively low Poisson's ratio (of approximately 0.15), its effect is usually ignored.

5.4.3 Torsion

The term 'torsion' as applied to cellular decks describes the shear forces and deformation induced by twisting the deck as in Fig. 5.3(d) without the effects of distortion of the cross-section in Fig. 5.3(e).

When a cellular deck is twisted bodily, there is a network of shear flows round the slabs and down and up the webs as shown in Fig. 5.10(a). Most of the shear flow passes round the perimeter slabs and webs, but some short-cuts through intermediate webs. When a grillage is twisted in a similar fashion, the forces crossing a transverse cross-section are as illustrated in Fig. 5.10(b). The total torque on a cross-section is made up partly from the torques in the longitudinal members and partly from the opposed shear forces on the two sides of the deck. These shear forces are in equilibrium with the torques in the transverse members as shown in Fig. 5.11. The system of forces in Fig. 5.10(b) is in fact very similar to that in the cellular deck of Fig. 5.10(a). By cutting the slabs between the webs as in Fig. 5.12 we can see that the grillage

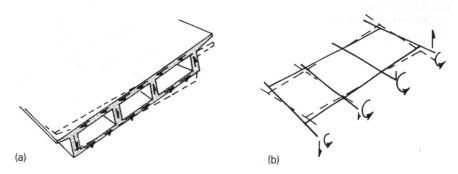

Fig. 5.10 Torsion forces on cross-section of twisted deck and grillage: (a) shear flows in deck and (b) torques and shear forces in grillage.

Fig. 5.11 Equilibrium of transverse torque and shear force in edge grillage member.

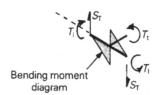

torques represent the torques in the deck due to the opposed shear flows in the top and bottom slabs, while the grillage shear force S_T represents the shear flow in the webs.

The torsional stiffness of a longitudinal or transverse grillage member is equal to that of the top and bottom slabs represented by the member. Their torsion constant is the same as that of two layers of similar thickness in a solid slab, giving

$$c = 2(h'^2 d' + h''^2 d'') = \frac{2h^2 d' d''}{(d' + d'')} \quad \text{per unit width of cell.} \tag{5.5}$$

This constant is equal to half the St Venant torsion constant (see Section 2.4.3) of a wide box beam per unit width. Just as in Chapter 3, the sum of the torsion stiffnesses of the longitudinal members of the grillage is only half the St Venant torsion stiffness of the section if it is treated as a beam. This reflects the fact that when a grillage is twisted the longitudinal member torques are only providing half of the total torque on the cross-section; the other half is provided by the opposed vertical shear forces on opposite sides of the deck. When the deck is twisted by differing amounts across the cross-section, as shown in Fig. 5.13, the relationship between slab and beam torsion constants is no longer relevant. However, the statical equivalence of the grillage and cellular deck force systems still holds.

The webs experience shear deformation due to the shear flows, and hence the longitudinal grillage members should be given shear areas equal to the cross-sectional areas of the webs.

Fig. 5.12 Statically equivalent torsion forces in deck and grillage.

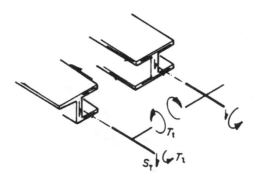

Fig. 5.13 Non-uniform torsion of (a) deck and (b) grillage.

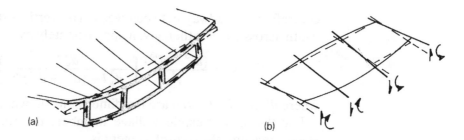

5.4.4 Distortion

Distortion of cells occurs as shown in Fig. 5.3(e) when the cells have few or no transverse diaphragms or internal bracing, so that a vertical shear force across a cell causes the slabs and webs to flex independently out-of-plane. This pattern of deformation is very similar to that of a Vierendeel truss of elevation similar to the cross-section of the deck. Although such behaviour cannot be reproduced precisely in a flat grillage, an approximation to this behaviour can be introduced by giving the transverse grillage members a low shear stiffness. The stiffness is chosen so that when the grillage member and cell are subjected to the same shear force, they experience similar distortions as in Fig. 5.14. One error in this analogy is that for the grillage beam the shear force is solely proportional to the shear displacement, while in the cell the shear force is to some extent dependent on the continuity of moments from flexure of the slabs in the adjacent cells. Fortunately it is found that the effect of this difference on the overall structural behaviour is small.

To determine the equivalent shear area of a transverse grillage member we must determine the relationship between vertical shear across a cell and the effective shear displacement w_S shown in Fig. 5.14. Precise frame analysis of a cell with differing thicknesses of top and bottom slabs and webs produces equations with unmanageable complexities. A convenient approximation is obtained by assuming that the shear force is shared between top and bottom slabs in proportion to their individual flexural stiffnesses and that there are points of

Fig. 5.14 Cell distortion and equivalent shear deformation of grillage member: (a) cell distortion and (b) shear deformation.

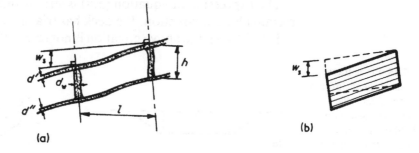

116 Multicellular decks

contraflexure midway between webs. The vertical shear force per unit width across a cell is then given approximately by

$$s \simeq \frac{(d'^3 + d''^3)}{l^3} \left[\frac{d_w^3 l}{d_w^3 l + (d'^3 + d''^3) h} \right] E w_S \qquad (5.6)$$

where d', d'', d_w, l and h are the dimensions shown in Fig. 5.14.

For the shear-flexible grillage member, the relationship between shear force and shear displacement is

$$s = \frac{a_S G w_S}{l} \quad \text{per unit width of member} \qquad (5.7)$$

where a_S is the equivalent shear area of the member.

Equating the stiffnesses of equations (5.6) and (5.7) we obtain the following expression for equivalent shear area of the grillage members

$$a_S = \frac{(d'^3 + d''^3)}{l^2} \left[\frac{d_w^3 l}{d_w^3 l + (d'^3 + d''^3) h} \right] \frac{E}{G} \quad \text{per unit width} \qquad (5.8)$$

If the webs in the deck are much closer together than it is practicable to place longitudinal grillage members, the transverse a_S is still calculated using the actual cell and web dimensions.

Equation (5.8) relates to a multicell deck. If it is used for a single-cell box-girder the terms $d_w^3 l$ should be replaced by $2d_w^2 l$ because the stiffness of the webs is not shared between two cells. However a plane frame analysis is likely to be more accurate and is straightforward as explained below and in Section 6.7.

Transverse members which represent cell distortion should not be subjected to direct loads, since the local bending will not be correctly modelled. Similarly nominal members should not be placed between widely spaced webs to simplify load description, as is advocated in Chapter 4 for beam-and-slab decks. Loads should be applied to a shear-flexible grillage at joints along longitudinal members at webs. Local bending of the top slab should be analysed with a separate grillage analysis for a slab strip with fixed edges, as described in Section 4.6. The edge reactions from the slab analysis then form the loads for the shear-flexible grillage.

The expression of equation (5.8) is strictly only applicable to cells of rectangular cross-section. If a deck has triangular or trapezoidal cells as in Fig. 5.15(a), the above equation is not relevant and the shear stiffness

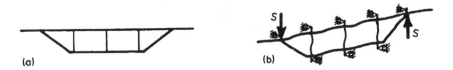

Fig. 5.15 Plane frame analysis of shear stiffness of trapezoidal cells: (a) deck cross-section and (b) plane frame subjected to distortion.

must be derived from a frame analysis of a frame of the same shape and dimensions as unit length of the deck. For complicated cross-sections this is most conveniently done using a computer plane frame analysis. The frame is supported as in Fig. 5.15(b) so that it cannot rotate and is subjected to distortional shear forces s. The shear stiffness of each cell is then s divided by the relative vertical movement across the cell. By equating the stiffness to $(a_S G/l)$ in equation (5.7), we obtain the equivalent shear area of the grillage member across the cell.

The transverse shear area of decks with cylindrical voids, as Fig. 5.1(d), can be calculated approximately by notionally replacing the cylindrical voids by square section voids of the same cross-sectional area. Equation (5.8) is then applied using the dimensions of the deck with square voids. Since the square voids are considerably more flexible than the cylinders, the calculated shear stiffness is an underestimate. An alternative method is to make a simple model of the deck cross-section such as the foam plastic model in Fig. 5.16(a). The deflections of this model due to distortion can be found by comparing the deflections with those of a solid-section beam with the same moment of inertia (as at the centre line of the voids). Fig. 5.16(b) shows the distortion of a Vierendeel beam with voids of the same area. To minimize the effects of creep, the beams should be loaded and inspected at precisely the same time intervals.

Where a transverse grillage member represents part of a cell with a diaphragm, the equivalent shear area is much larger and includes the cross-sectional area of the diaphragm.

Cell distortion is not just resisted by out-of-plane flexure of the top and bottom slabs and the webs but also by in-plane bending and shear of these plate elements. Fig. 5.17(a) shows the cross-section and elevation of a cell whose faces have experienced in-plane bending, while Fig. 5.17(b) shows a cell distorted by in-plane shear of the cell. The shear flows that cause the distortion of Fig. 5.17(b) can be precisely the same as those that produce torsion in Fig. 5.17(c). If the section is prevented from distorting, torsion occurs as in Fig. 5.17(c) without change of cross-sectional shape but with warping of the cross-section (i.e. alternate

Fig. 5.16 Foam plastic cellular and solid beams flexed back to back to compare deflections under identical loads.

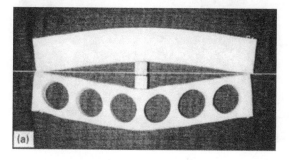
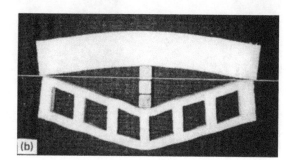

118 Multicellular decks

Fig. 5.17 Cell distortion and torsion with in-plane deformation of plates: (a) distortion with in-plane bending of plates; (b) distortion with in-plane shear of plates; and (c) torsion with in-plane shear of plates.

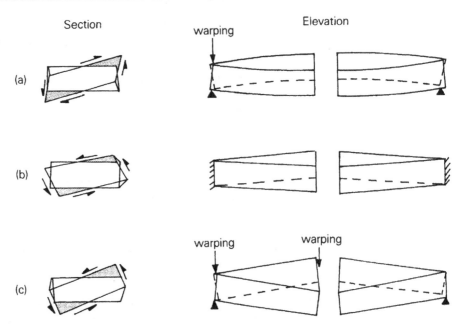

Fig. 5.18 Foam plastic models of part of box exhibiting: (a) shear distortion; (b) twisting distortion with warping; and (c) torsion.

longitudinal movements in the corners); if the deck has a higher stiffness against warping than distortion, the shear deformation of the faces causes the cell to distort as in Fig. 5.17(b).

The differences between distortion and torsion of a cellular structure are illustrated by the plastic models of part of a box in Fig. 5.18. Readers are recommended to make similar models for themselves, or use a cardboard box, and feel the different stiffnesses for themselves. The

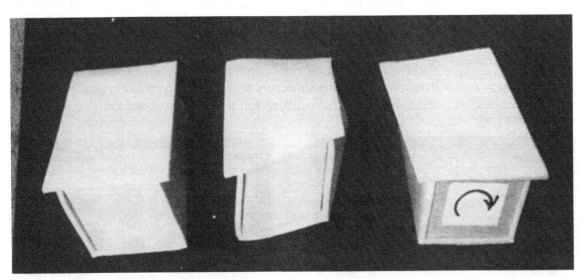

model on the left is being subjected to distortion, as described in the preceding paragraphs, with slabs and webs flexing and plane sections of the box remaining plane (i.e. no warping). The distortion of the model was created by inserting lozenge-shaped diaphragms into the ends of the model with both ends distorted in the same direction. The box structure is flexible for this mode of distortion and little force is needed to insert the diaphragms. The box structure of the middle model is being subjected to distortion of a totally different kind by inserting the lozenge diaphragms with opposite orientations. Half-way along the model the cross-section is square. This distortion is accompanied by warping so that plane sections have warped substantially out-of-plane. The top slab has rotated anticlockwise without changing shape, while the bottom slab has rotated clockwise. The webs have also rotated in opposite directions. Again relatively little force is needed to insert the lozenge diaphragms indicating that the short length of box is flexible in this mode of distortion. A wide cellular deck resists this type of distortion by the large in-plane stiffnesses of the slabs which resist their in-plane movements. However a narrow box-girder can be flexible in this mode of distortion. The model on the left is being subjected to torsion. The diaphragms at its ends have been twisted relative to each other (and held in twisted position by a piece of wood through the middle of the model). It is found that a large torque is needed to produce a discernible twist in this manner so indicating that the stiffness of this model against torsion is several orders of magnitude greater than against distortion. Large torsion shear flows are required to shear each face of the box in its plane.

A shear-flexible grillage model reproduces the distortion flexibility of the left model by the shear area of transverse members given by equation (5.8). The grillage reproduces the high torsion stiffness of the right model by the high torsion constant given to longitudinal and transverse members with equation (5.5). The distortion of the middle model is reproduced to a certain extent in a grillage by the differences in deflected shape between adjacent longitudinal beams. However the shear-flexible grillage does not reproduce the low distortional stiffness of a narrow box-girder, and this requires the more detailed methods reported in Chapter 6.

The design of a cellular structure involves a compromise between torsional stiffness, which is beneficial to efficient load distribution, and distortional flexibility which helps the structure spread the loads between supports. A box structure which is very stiff against distortion and torsion may not rest comfortably on bearings under adjacent webs, and the distribution of reactions between webs could be very sensitive to compressibility and differential settlement of bearings, as is discussed in Chapter 14. If two bearings are close the whole reaction may go into one bearing.

A box section is unlikely to have sufficient stiffness to control distortion deformations at a concentrated load, such as at a support reaction, or sufficient strength to transfer shear loads between webs at bearings. A diaphragm, or additional framing, is probably required at such a point. The action of diaphragms is discussed in Chapter 6. A diaphragm provides a diagonal tying/propping action which reacts with the distortion shear flows shown in Fig. 6.2(f).

5.4.5 Stiffened steel plate structures

The section properties in Sections 5.4.1 to 5.4.4 have all been expressed in terms of thicknesses of slabs and webs. Similar equations can be derived for the stiffnesses of stiffened steel plates in steel box-girders. However the effective properties of a stiffened plate are different under longitudinal compression, transverse compression, flexure and shear.

The longitudinal bending inertia for a stiffened box-girder is calculated from the effective areas of the plates and stiffeners under longitudinal compression. The transverse bending inertia of equation (5.4) is calculated from the effective areas of the plates and framing under transverse compression, while the torsion constant of equation (5.5) is calculated from the effective stiffnesses of the stiffened plates under shear (from torsion). The effective shear area for transverse distortion in equation (5.8) is based on the effective stiffnesses of the plates and frames under transverse flexure. This equation can be expressed in the form

$$a_s = \frac{12(i' + i'')}{l^2} \left[\frac{i_w l}{i_w l + (i' + i'')h} \right] \frac{E}{G} \quad (5.9)$$

where i', i'' and i_w are the effective flexural inertias per unit length of the top plate, bottom plate and webs, each with framing. These inertias can be expressed in terms of equivalent flexural rigidities D', D'', D_w defined using equation (3.5). The effective shear area can also be obtained from a plane frame model using these inertias.

Equation (5.9), like equation (5.8), relates to a multicell structure. If it is used for a single-cell box-girder the equation takes the form

$$a_s = \frac{12(i' + i'')}{l^2} \left[\frac{2i_w l}{2i_w l + (i' + i'')h} \right] \frac{E}{G}. \quad (5.10)$$

However a plane frame analysis, as demonstrated in Section 6.7, is likely to be more accurate and just as quick. If the framing is not continuous around the box a plane frame analysis is almost definitely required.

Fig. 5.19 Twin-cell concrete box girder bridge over M42, England; designed by Gifford & Partners. See Fig. 5.20. Photograph E. C. Hambly.

122 Multicellular decks

5.5 SECTION PROPERTIES OF GRILLAGE MEMBERS

This section summarizes the findings of the preceding section and demonstrates calculated examples of grillage section properties for three different decks. The notation for the dimensions of the cells, slabs and webs is defined in Fig. 5.14(a).

5.5.1 Grillage for three-span twin-cell box-girder deck

Figure 5.20 gives details of a three-span twin-cell box-girder with supports at 21° skew. The grillage mesh is chosen with three 'structural' longitudinal members 2, 3 and 4 coincident with the webs. Two 'nominal' members 1 and 5 are located along the edges of the cantilevers. Transverse members representing the top and bottom slabs are orthogonal to the longitudinal members. Along the spans their spacing is approximately one quarter of the distance between points of contraflexure, but over the intermediate supports the spacing is shorter to give greater detail near peaks in the bending moment diagrams. At

Fig. 5.20 Grillage for three span twin-cell concrete box-girder deck: (a) deck section; (b) grillage section; (c) deck longitudinal section; and (d) grillage mesh.

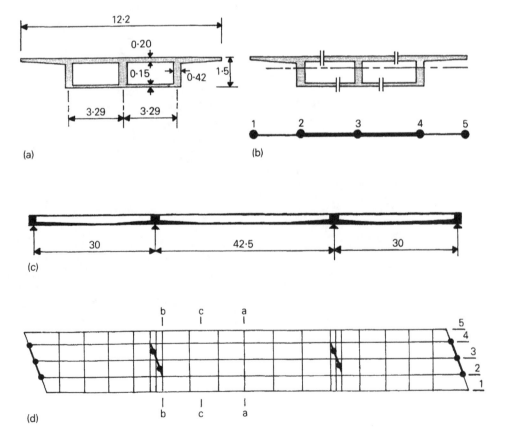

the ends the skew members represent the slabs and the diaphragm, while at the internal supports the skew member represents just the solid diaphragm without flanges.

The 'structural' longitudinal members 2, 3 and 4 have the moments of inertia of the 'I-beams' obtained by cutting the deck as in Fig. 5.20(b) so that the centroid of each 'I-beam' is on the principal axis of the deck. In this case each beam includes one third of the top slab and one third of the bottom slab. The moment of inertia of each is one third of the total moment of inertia of the deck

$$I_2 = I_3 = I_4 = \frac{1.54}{3} = 0.51 \, \text{m}^4.$$

The above calculation has ignored the reduction of effective flange widths of the sections due to shear lag. This is significant in this deck, particularly near the intermediate supports, and a simple correction described in Section 8.2 is strictly necessary.

The torsion constant per unit width is given by equation (5.5)

$$c = \frac{2h^2 d' d''}{(d' + d'')} \quad \text{per unit width}$$

$$c = \frac{2 \times 1.325^2 \times 0.2 \times 0.15}{(0.2 + 0.15)} = 0.30 \, \text{m}^4 \, \text{m}^{-1}.$$

The widths of cell in members 2, 3 and 4 are 3.29/2, 3.29 and 3.29/2, respectively. Hence their torsion constants are

$$C_2 = C_4 = \frac{3.29}{2} \times 0.30 = 0.49 \, \text{m}^4 \quad C_3 = 3.29 \times 0.30 = 0.99 \, \text{m}^4.$$

The shear areas of members 2, 3 and 4 are equal to the areas of the webs:

$$A_{S2} = A_{S3} = A_{S4} = 0.42 \times 1.325 = 0.56 \, \text{m}^2.$$

Near each support the bottom slab of the deck is thickened. In these regions, the properties of each grillage member are calculated in the same manner as above for the section midway along its length.

The 'nominal' edge members have the section properties of half the cantilever

$$I_1 = I_5 = \frac{bd'^3}{12} = \frac{2.81}{2} \times \frac{0.2^3}{12} = 0.000\,94 \, \text{m}^4$$

$$C_1 = C_5 = \frac{bd'^3}{6} = \frac{2.81}{2} \times \frac{0.2^3}{6} = 0.0019 \, \text{m}^4$$

$$A_{S1} = A_{S5} = bd' = \frac{2.81}{2} \times 0.2 = 0.28 \, \text{m}^2.$$

The transverse members representing cells have section properties given by equations (5.4), (5.5) and (5.8):

$$i_{23} = \frac{h^2 d' d''}{(d' + d'')} \text{ per unit width}$$

$$= \frac{1.325^2 \times 0.2 \times 0.15}{(0.2 + 0.15)} = 0.15 \, \text{m}^4 \, \text{m}^{-1} \quad \text{From equation (5.4)}$$

$$c_{23} = \frac{2h^2 d' d''}{(d' + d'')} \text{ per unit width}$$

$$= 2 \times 0.15 = 0.30 \, \text{m}^4 \, \text{m}^{-1} \quad \text{From equation (5.5)}$$

$$a_{S23} = \frac{(d'^3 + d''^3)}{l^2} \left[\frac{d_w^3 l}{d_w^3 l + (d'^3 + d''^3) h} \right] \frac{E}{G} \text{ per unit width}$$

$$= \frac{(0.2^3 + 0.15^3)}{3.29^2} \left[\frac{0.42^3 \times 3.29}{0.42^3 \times 3.29 + (0.2^3 + 0.15^3) 1.325} \right] 2.3$$

$$= 0.0024 \, \text{m}^2 \, \text{m}^{-1}.$$

Transverse members on the cantilever have the properties of the top slab

$$i_{12} = \frac{d^3}{12} \text{ per unit width} = \frac{0.2^3}{12} = 0.000\,67 \, \text{m}^4 \, \text{m}^{-1}$$

$$c_{12} = \frac{d^3}{6} \text{ per unit width} = \frac{0.2^3}{6} = 0.001\,34 \, \text{m}^4 \, \text{m}^{-1}$$

$$a_{S12} = d \text{ per unit width} = 0.2 \, \text{m}^2 \, \text{m}^{-1}.$$

The skew members representing the internal diaphragm (here 1.5 m wide) have the properties of the solid section. For derivation of the torsion constant see Section 2.4.2.

$$I = \frac{1.5 \times 1.325^3}{12} = 0.29 \, \text{m}^4$$

$$C = \frac{3 \times 1.5^3 \times 1.325^3}{10(1.5^2 + 1.325^2)} = 0.59 \, \text{m}^4$$

$$A_S = 1.5 \times 1.325 = 2.0 \, \text{m}^2.$$

5.5.2 Grillage for wide multicellular deck

Figure 5.21 gives details of a three-span multicellular deck at high skew angle. The section properties of the grillage members are derived as in the preceding example except that it is inconvenient to split the deck up into beams with individual centroids precisely on the principal axis of the deck. Consequently, the deck is notionally cut midway between webs as shown in (b). The centroids of internal 'structural' members 3, 4, etc. are virtually coincident with the principal axis of the deck. Edge 'structural' member 2 has its centroid at a higher level but, like the other members, its grillage moment of inertia is calculated about the principal axis of the deck.

5.5.3 Grillage for cellular deck with inclined webs

Figure 5.22 gives details of part of a multispan four-cell deck with inclined edge webs and haunches at the supports. It is impracticable to split up the deck into longitudinal members with centroids on the principal axis of the deck, hence the deck is notionally cut as in (a) into five 'structural' members with inertias calculated about the principal axis of the deck. There are no 'nominal' edge members in order to permit economy in the size of the grillage.

There is no clear-cut equivalence of grillage torsion stiffness for the non-rectangular cells. However, sensible results are obtained if equation (5.6) is used with h equal to the average height of the cell.

Fig. 5.21 Grillage for three-span high-skew multicell concrete deck: (a) part deck section; (b) grillage section; and (c) grillage mesh.

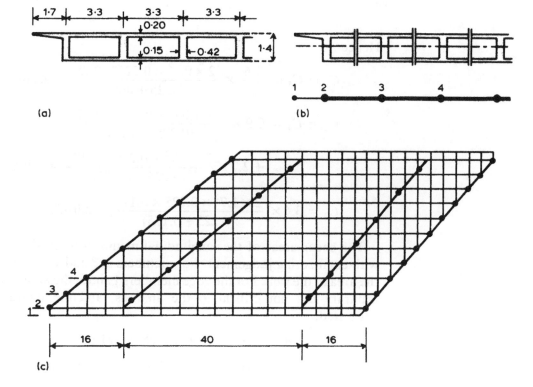

Fig. 5.22 Part of grillage for multispan deck with haunches and inclined webs: (a) deck section; (b) grillage section; (c) deck longitudinal section; and (d) grillage mesh.

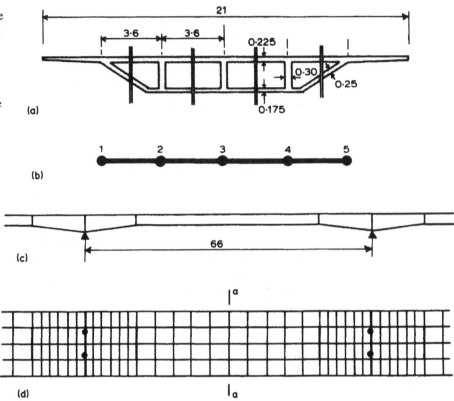

Hence

$$C_1 = C_5 = \frac{3.6}{2} \times \frac{2 \times 0.5^2 \times 0.225 \times 0.25}{(0.225 + 0.25)} = 0.1 \, \text{m}^4$$

$$C_2 = C_4 = 0.9 \times \frac{2 \times 1.35^2 \times 0.225 \times 0.25}{(0.225 + 0.25)}$$

$$+ 2.7 \times \frac{2 \times 1.7^2 \times 0.225 \times 0.175}{(0.225 + 0.175)} = 1.9 \, \text{m}^4$$

$$C_3 = 3.6 \times \frac{2 \times 1.7^2 \times 0.225 \times 0.175}{(0.225 + 0.175)} = 2.0 \, \text{m}^4.$$

The shear areas of the longitudinal members must also be derived with engineering judgement. For members 2, 3 and 4, the shear area is the area of the web. For the edge members 1 and 5, the area is somewhat arbitrary but not critical to the analysis, and the figure below is suggested.

$$A_{S1} = A_{S5} = 0.9 \times 0.25 = 0.21 \, \text{m}^2$$

$$A_{S2} = A_{S3} = A_{S4} = 1.7 \times 0.3 = 0.51 \, \text{m}^2.$$

The transverse grillage members also require special consideration and judgement related to the particular geometry of the cross-section. It is suggested that the moment of inertia and torsion constant can generally be calculated with equations (5.4) and (5.5) using the average value of h across the cell. The shear areas must be derived from a plane frame analysis as described in Section 5.4.4. Such an analysis of the cross-section of Fig. 5.22(a) gave the following values:

$$\text{edge cells} \quad A_{S12} = 0.05 \text{ m}^2 \text{ m}^{-1}$$
$$\text{internal cells} \quad A_{S23} = 0.005 \text{ m}^2 \text{ m}^{-1}.$$

5.6 LOAD APPLICATION

The webs of many cellular and box-girder decks are spaced further apart than the width of the carriageway lanes. Consequently it is possible for a whole lane of loading to lie between grillage members. These loads can be applied to the grillage joints on each side by statical distribution. Since the deck has high transverse bending and longitudinal torsional stiffnesses, its behaviour under these statically distributed loads is virtually the same as if the loads had been applied more correctly as the fixed edge shear forces and moments at the edges of the top slab. This contrasts with the behaviour of a spaced beam-and-slab deck described in Section 4.6 where it was shown that with its low transverse bending and longitudinal torsional stiffnesses, the beam-and-slab deck deflects in different ways when a load is placed between and when it is statically distributed on to the beams.

The grillage of a cellular deck, like that of a beam-and-slab deck, only gives the force systems in the deck due to deformation of the structure as a whole. It does not give an indication of the local moments and shear forces due to the concentration of load between grillage members. These moments and shear forces must be derived independently using a separate local grillage analysis, or the charts of Pucher [3], as described for beam-and-slab decks in Section 4.6, and added to those from the load distribution in the grillage.

5.7 INTERPRETATION OF OUTPUT

The output of the grillage of a cellular deck should be interpreted with as much care as the calculation of section properties. A considerable amount of valuable detailed information about the forces in the cells can be derived from the grillage by reapplying the principle of local static equivalence of forces in cellular deck and grillage.

The following discussion is illustrated with examples of stresses, etc. calculated from the grillage analysis of the deck in Fig. 5.20. The load case for which they are relevant comprised full lane loading and footpath loading on one side of the main span.

5.7.1 Longitudinal bending

The bending moment diagrams for the three 'structural' longitudinal grillage members are shown in Fig. 5.23. The diagrams have saw teeth with large jumps in moment at the joints because of the transfer of the torsion in the transverse members at each joint to bending moments and shear forces in the longitudinal member as was shown in Fig. 5.11. The 'true' bending moment diagrams can be assumed to pass through the average values of the bending moment on the two sides of each joint as shown by the dashed lines in Fig. 5.23.

The longitudinal bending stresses at a section are calculated from these 'true' moments using the section properties of the I-beam represented by the grillage member. Figure 5.24(a) and (b) show the bending stresses calculated in this way for sections a–a and b–b of the deck. The bending stresses are shown constant across each I-beam without smoothing out the impossible jumps in stress at the notional cuts between I-beams and without correction for shear lag as described in Chapter 8.

The shear force S_M due to bending is the slope of the 'true' dashed bending moment diagram in Fig. 5.23. The shear flows in the web and flanges of the 'I-beams' are calculated from S_M using equation (5.2). Figure 5.25 shows these bending shear flows at cross-section c–c.

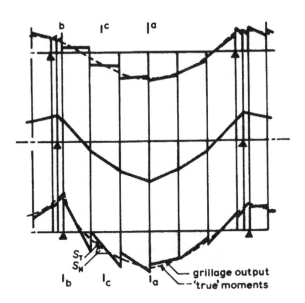

Fig. 5.23 Longitudinal moments in part of grillage.

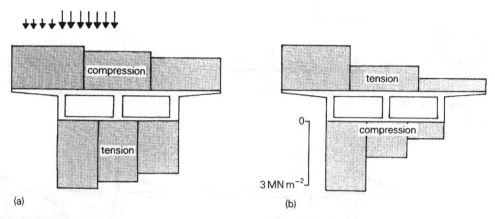

Fig. 5.24 Longitudinal bending stresses: (a) sagging at midspan a–a and (b) hogging at support b–b.

5.7.2 Transverse bending

The transverse bending moment in the grillage member is equivalent to the opposed transverse compression of the top slab and tension of the bottom slab (or vice versa) due to transverse flexure without distortion. In narrow decks it is usually very small compared to the longitudinal bending moment (except in the diaphragms). However, in wide decks it can be large especially near skew supports. Since these moments interact with the torsions in longitudinal grillage members, a grillage output transverse moment diagram has a saw tooth shape like the longitudinal moment diagram, and in a similar way the top and bottom slab stresses are calculated from the average moments on the two sides of each joint.

5.7.3 Slab flexure from distortion

The stiffness of the cell against independent flexure of the slabs during distortion is represented by the shear stiffness of transverse grillage members. For this reason the slab bending moments are derived from the shear force in the transverse grillage members. Figure 5.26(a) shows the

Fig. 5.25 Longitudinal bending shear flow at section c–c.

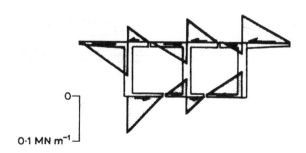

130 Multicellular decks

Fig. 5.26 Slab transverse moments at section a–a: (a) grillage transverse shear forces; (b) cell distortion moments; (c) cantilever and local moments; and (d) total moments.

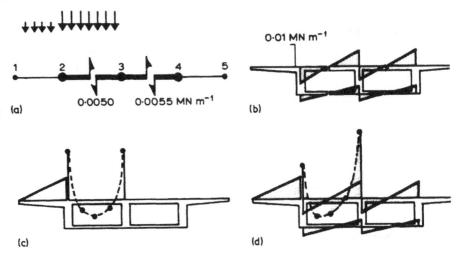

Fig. 5.27 Torsion shear flows at c–c calculated from average of transverse and longitudinal torques output by grillage: (a) torques in grillage members per unit width and (b) shear flows.

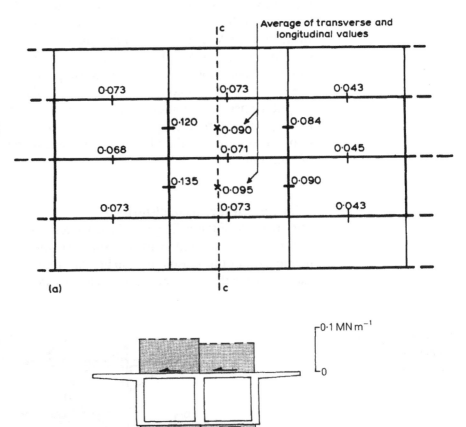

shear force in the transverse members on section a–a of Fig. 5.20(d). The fractions of this shear force carried by each of the top and bottom slabs are assumed to be proportional to the flexural stiffnesses of the slabs (here 0.7:0.3). Assuming also that points of contraflexure lie midway between the webs, the moment at each end of a slab is simply the shear force it carries times half the distance between webs. Hence we obtain the transverse moment diagram of Fig. 5.26(b) from the shear forces of (a). The transverse slab moment in the cantilever can be taken directly from the grillage output since this member is not representing a cell. Figure 5.26(c) shows the cantilever moment and the local moments under the knife-edge load above section a–a, while (d) shows the total moments obtained by adding (b) and (c). The local moments are derived from a local grillage analysis as described in Section 4.6.

5.7.4 Torsion shear flow

The torsion shear flows in the slabs must be calculated from the average torque per unit width of transverse and longitudinal grillage member as described in Section 5.4.4. Figure 5.27(a) shows the torques in grillage members per unit width of cell; the averages of the values in the two directions are also shown. By dividing these average torques per unit width of cell by the distance h between slab midplanes, we obtain the shear flows in the slabs shown in Fig. 5.27(b). On adding these to the bending shear flows in Fig. 5.25, we obtain the total shear flows shown in Fig. 5.28.

The shear forces output from the grillage for longitudinal members are the slopes of the saw teeth of the output moment diagram of Fig. 5.23. These shear forces combine the components S_M due to bending (the slope of the dashed 'true' moment diagram) and the components S_T due to torsion (the additional slope of the teeth caused by transverse torsion). Consequently, the grillage output shear force represents the total shear force in each web of the deck.

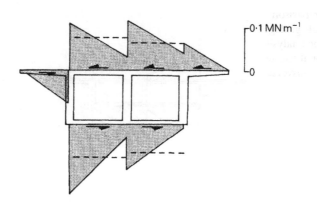

Fig. 5.28 Combined shear flow from torsion and bending.

Fig. 5.29 Comparison of longitudinal bending stresses from shear flexible grillage with finite strip analysis: (a) midspan and (b) over skew pier.

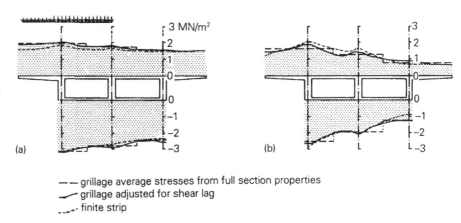

– – grillage average stresses from full section properties
—— grillage adjusted for shear lag
-·-·- finite strip

Since the torsional shear force S_T in longitudinal members results from equilibrium with torques in transverse members, it is in error if the transverse torque differs significantly from the average of the torques in the two directions. Although it is seldom necessary, a correction can be made to the torsional shear force by decreasing it or increasing it as appropriate in proportion to the excess of the transverse torque over the average torque.

5.8 COMPARISON WITH FINITE STRIP METHOD

Figure 5.29 and Fig. 5.30 present a comparison of the results from a shear-flexible grillage analysis for the deck of Fig. 5.20 with the results of a finite strip analysis. These results were reported in reference [4]. The finite strip method is described in Section 13.5 and is a type of finite element analysis which reproduces all the in-plane and out-of-plane stiffnesses of all the slab and web parts of the structure. The grillage

Fig. 5.30 Comparison of shear flexible grillage with finite strip analysis: (a) shear flows at quarter span and (b) transverse bending.

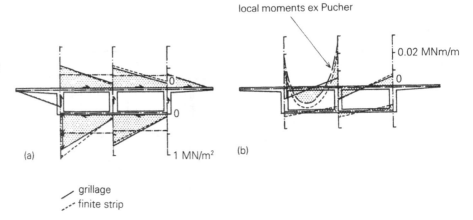

—— grillage
-·-·- finite strip

Fig. 5.31 High skew multicellular concrete slab deck Southampton, England; designed by Gifford & Partners. See Fig. 5.21. Photograph E. C. Hambly.

model was adjusted specially for this comparison to have exactly the same geometry as the finite strip analysis, which had to have a prismatic cross-section and right end supports. The grillage output was interpreted in the manner described in Section 5.7 with shear lag correction as described in Section 8.2. Figure 5.29(a) and (b) illustrate the distributions of bending stresses calculated for cross-section at midspan and at the skew pier when the centre span was loaded on one side. Figure 5.30(a) shows the shear flows around a cross-section at quarter span due to bending and torsion while Fig. 5.30(b) shows the transverse bending moments in the slabs at midspan. It is evident that all the results from the shear-flexible grillage are remarkably close to the results from the finite strip analysis. Another satisfactory comparison between a shear-flexible grillage and a finite strip analysis for a different structure is presented in Section 13.5.

REFERENCES

1. Sawko, F. (1968) 'Recent developments in the analysis of steel bridges using electronic computers', British Construction Steel Association Conference on Steel Bridges, London, pp. 39–48.
2. Hambly, E.C. and Pennells, E. (1975) 'Grillage analysis applied to cellular bridge decks', *Structural Engineer*, **53**, No.7, pp. 267–75.
3. Pucher, A. (1964) *Influence Surfaces of Elastic Plates*, Springer-Verlag, Vienna and New York.
4. Hambly, E.C. (1974) Contribution to discussion on 'Concrete Box-Girder Bridges' by B.I. Maisel, R.E. Rowe and R.A. Swann, *Structural Engineer*, **52**, No.7, pp. 257–8.

6 Box-girder decks

6.1 DISTORTION OF SINGLE-CELL BOX-GIRDER

The distortion of a single-cell box-girder can be substantially greater than that of a multicell structure because the top and bottom flanges are able to bend in-plane sideways. Figure 6.1(a) illustrates the distortion of a box-girder. The top flange is moving to the right, the bottom to the left, the right web is moving up and the left web is moving down. (A photograph of distortion of plastic models of short lengths of box is shown in Fig. 5.18.) Figure 6.1(b) shows the transverse bending moments due to out-of-plane flexure of the plates, while Fig. 6.1(c) shows the longitudinal stresses due to in-plane bending.

Figure 6.2 shows how the distortion forces develop in the structure. The eccentric load in Fig. 6.2(a) can be thought of as an antisymmetric component in (b) and a symmetric component in (c). The symmetric component causes vertical bending of the whole box-girder, as discussed in Chapter 2, and is not discussed further. The antisymmetric component cannot be equated directly to torsion on the box because pure torsion involves a system of shear flows round the cell as shown in Fig. 6.2(e). The antisymmetric load in Fig. 6.2(b) which is redrawn in

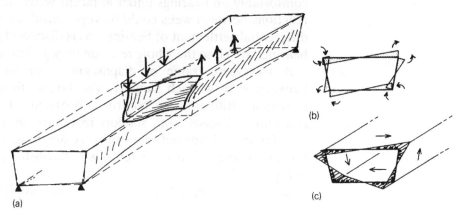

Fig. 6.1 (a) Distortion of box-girder deck; (b) out-of-plane bending moments; and (c) in-plane bending (warping) stresses.

Fig. 6.2 (a) Eccentric load; (b) antisymmetric component; (c) symmetric component; (d) = (b); (e) pure torsion shear flows; and (f) distortion shear flows.

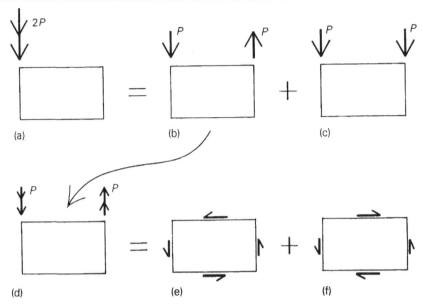

(d) is equivalent to the combination of the pure torsion shear flows in (e) and the distortion shear flows in (f). The torque involved in the pure torsion in (e) is equal to the torque of the antisymmetric loading (d). The distortion shear flows in (f) balance each other and have no net resultant but at the same time they cause distortion of the cell as shown in Fig. 6.1. A box-girder is very stiff in pure torsion and most of the twist of the deck is due to distortion, unless the box is braced with diaphragms or cross-bracing.

Transverse bracing or framing provide a very effective method for stiffening a box-girder against distortion. The amount of stiffening that is appropriate involves a compromise between torsional stiffness, which is beneficial to efficient load distribution, and distortional flexibility which helps the structure spread the loads between supports. A box structure which is very stiff against distortion and torsion may not rest comfortably on bearings under adjacent webs, and the distribution of reactions between webs could be very sensitive to compressibility and differential settlement of bearings, as is discussed in Chapter 14. If the bearings are close the whole reaction may go into one bearing.

A transverse frame or diaphragm is almost definitely required between webs at each support. The box section is unlikely to have sufficient stiffness to control distortion deformations at the concentrated loads from support reactions, or sufficient strength to transfer shear loads between webs. A diaphragm provides a diagonal tying/propping action which reacts with the distortion shear flows shown in Fig. 6.2(f).

Distortion of curved box-girders is more complicated because torsion and bending interact along the span, as discussed in Chapter 9, and all loads contribute to distortional deflections and stresses.

6.2 METHODS OF CALCULATION

Several very complicated mathematical methods have been published for analysing distortion of box-girders; a review is provided by Nakai and Yoo [1]. Hand methods of calculation can be used for structures of relatively simple geometry, as explained in references [2]–[7]. The BEF method ('beam-on-elastic-foundations'), in references [2]–[4], takes advantage of the similarity of the equations for distortion of a box-girder with the equations for a beam on elastic foundations (BEF). The method is demonstrated in Section 6.3 below. References [3] and [5] provide relatively simple equations for the analysis of concrete box-girders with no diaphragms or cross-bracing. Tables of non-dimensional parameters for longitudinal warping and transverse bending stresses are provided in references [1] and [6].

The mathematical complexity can be avoided if the structure is analysed with a three-dimensional computer model of finite elements or a space frame, which are demonstrated in Section 6.4. The computer models do not need to be complicated. Global analyses can be undertaken with various grillage simulations. Sections 6.5 and 6.6 illustrate grillage analyses of unbraced and braced box-girder decks and of a multiple box-girder structure. Section 6.7 demonstrates a different grillage simulation, using shear flexibility as explained in Chapter 5, for a braced multispan box-girder.

6.3 BEF ANALYSIS OF BOX-GIRDER

This demonstration of the BEF method follows Wright *et al.* [2] in order to illustrate the influence of cross-bracing. Two design charts are reproduced from reference [2] with some of the equations.

Figure 6.3 illustrates a single-span composite steel box-girder bridge supporting a vehicle of weight $2P$ over one web near midspan. The following calculation examines the distortion caused by the antisymmetric component of the load, shown in Fig. 6.3(e), which consists of the up and down loads of magnitude P at the two webs. The symmetric component in Fig. 6.3(d) causes ordinary bending and is not considered here.

The structure has a span of 30 m. The cross-section has a concrete top flange 6.6 m wide and 0.2 m thick and the box dimensions are: $a = 3.2$ m, $b = 2.8$ m, $c = 1.5$ m, $d = 3.4$ m, $h = 1.5$ m. The bottom flange is 0.020 m thick and the webs 0.010 m thick. The structure is first considered

138 Box-girder decks

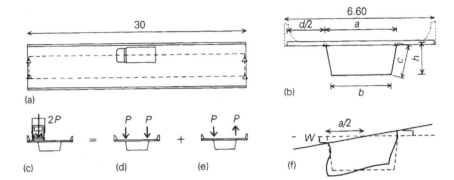

Fig. 6.3 Single span composite steel box-girder deck: (a) plan; (b) section; (c), (d) and (e) components of load; and (f) distortion deflections.

without any cross-bracing or diaphragms, except at supports. It is assumed that Young's modulus $E = 200\,000$ MPa for steel and $30\,000$ MPa for concrete. Poisson's ratio v is assumed to be 0.25 for steel and concrete. All section properties for concrete are converted below to the equivalent for steel using a modular ratio $m = 0.15$.

The top flange, bottom flange, and webs have the following thicknesses and flexural rigidities $D = Et^3/(12(1 - v^2))$:

top flange: $\quad t_a = 0.15 \times 0.2 = 0.03$ m

$$D_a = \frac{0.15 \times 200\,000 \times 0.2^3}{12(1 - 0.25^2)} = 21.3 \text{ MNm}$$

bottom flange: $\quad t_b = 0.020$ m

$$D_b = \frac{200\,000 \times 0.02^3}{12(1 - 0.25^2)} = 0.142 \text{ MNm}$$

webs: $\quad t_c = 0.010$ m

$$D_c = \frac{200\,000 \times 0.01^3}{12(1 - 0.25^2)} = 0.018 \text{ MNm}$$

The overall cross-section of the box-girder has equivalent moment of inertia I_c of

$$I_c = 0.11 \text{ m}^4.$$

The depth of the neutral axis below the centre of the top flange is $\bar{y}$, with

$$\bar{y} = 0.37 \text{ m} \quad h - \bar{y} = 1.13 \text{ m}.$$

Reference [2] equation (10) (reproduced as equation (6.1) below) indicates that the out-of-plane shear in the bottom flange per unit torsional load is

$$v = \frac{\frac{1}{D_c}[(2a + b)abc] + \frac{1}{D_a}[ba^3]}{(a + b)\left[\frac{a^3}{D_a} + \frac{2c}{D_c}(a^2 + ab + b^2) + \frac{b^3}{D_b}\right]} \quad (6.1)$$

which here gives

$$v = \frac{\frac{1}{0.018}[(2 \times 3.2 + 2.8)3.2 \times 2.8 \times 1.5] + \frac{1}{21.3}[2.8 \times 3.2^3]}{(3.2 + 2.8)\left[\frac{3.2^3}{21.3} + \frac{2 \times 1.5}{0.018}(3.2^2 + 3.2 \times 2.8 + 2.8^2) + \frac{2.8^3}{0.142}\right]}$$

$= 0.25$.

Reference [2] equation (11) (reproduced as equation (6.2) below) indicates that the vertical deflection of one web per unit torsional load is

$$\delta_1 = \frac{ab}{24(a+b)} \left\{ \frac{c}{D_c}\left[\frac{2ab}{a+b} - v(2a+b)\right] + \frac{a^2}{D_a}\left[\frac{b}{a+b} - v\right]\right\} \quad (6.2)$$

which here is

$$\delta_1 = \frac{3.2 \times 2.8}{24(3.2 + 2.8)} \left\{ \frac{1.5}{0.018}\left[\frac{2 \times 3.2 \times 2.8}{3.2 + 2.8} - 0.25(2 \times 3.2 + 2.8)\right]\right.$$
$$\left. + \frac{3.2^2}{21.3}\left[\frac{2.8}{3.2 + 2.8} - 0.25\right]\right\}$$

$= 3.6 \, \text{m}^2/\text{MN}$.

Reference [2] equation (34) (equation (6.3) below) defines a BEF parameter

$$\beta = \left\{\frac{1}{4EI_b \delta_1}\right\}^{0.25} = \left\{\frac{1}{EI_c \delta_1}\right\}^{0.25} \quad (6.3)$$

where I_b is the BEF moment of inertia, which is approximately $0.25 I_c$.

Here

$$\beta = \left\{\frac{1}{200\,000 \times 0.11 \times 3.6}\right\}^{0.25} = 0.06 \, \text{m}^{-1}.$$

Reference [2] equation (35) (equation (6.4) below) relates the distortion deflection W in Fig. 6.3(f) to the concentrated torsional load P with the dimensionless deflection term w, in the form

$$w = 8EI_b \beta^3 \frac{W}{P}. \quad (6.4)$$

With $I_b = 0.25 I_c$ equation (6.4) can be rearranged to give

$$\frac{W}{P} = \frac{w}{8EI_b \beta^3} = \frac{w}{2EI_c \beta^3}$$

which here gives

$$\frac{W}{P} = \frac{w}{2 \times 200\,000 \times 0.11 \times 0.06^3} = 0.1\,w.$$

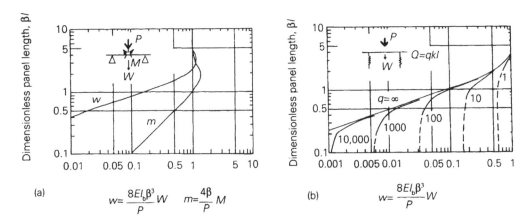

Fig. 6.4 BEF charts for distortion deflections of box girders: (a) simply supported BEF and (b) continuous BEF. (From Wright, Abdel-Samad and Robinson [2]).

Reference [2] provides charts of w versus the dimensionless panel length βl for a simply supported BEF and a continuous BEF, and these charts are reproduced in Fig. 6.4(a) and (b). The chart in (a) for a simply supported BEF is used when the box has no cross-bracing or diaphragms between supports, while the chart in (b) for a continuous BEF is used when the box has bracing at points along its length.

In this case without cross-bracing the distorting panel length l is equal to the span, giving $l = 30$, and the dimensionless panel length is

$$\beta l = 0.06 \times 30 = 1.8.$$

From the chart of Fig. 6.4(a) we find that w is about 0.6 when $\beta l = 1.8$. Hence from equation (6.4) above

$$W/P = 0.1 \times 0.6 = 0.06 \text{ m/MN}.$$

If the load in Fig. 6.3 represents a vehicle of weight $2P = 1$ MN (100 tonne) standing over one web, then the antisymmetric component of

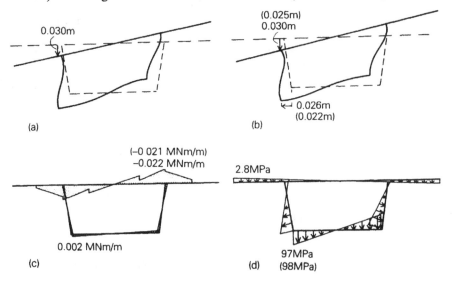

Fig. 6.5 (a) Distortion at midspan of box-girder calculated by BEF method; (b) distortion calculated by space frame; (c) transverse bending moments from space frame; and (d) longitudinal warping stresses from space frame. Results from finite element analysis in brackets.

load is $P = 0.5$ MN. Hence the distortional deflection W

$$W = 0.06 \times 0.5 = 0.030 \text{ m}.$$

This deflection is shown in Fig. 6.5(a).

Now consider the effect of stiffening the cross-section with cross-bracing at 7.5 m centres. Assume that each brace has area $A_b = 0.002$ m^2, length $L_b = 3.3$ m and Young's modulus $E_b = 200\,000$ MN/m^2. The panel length is now $l = 7.5$ m so that $\beta l = 0.06 \times 7.5 = 0.45$.

Reference [2] equation (12) (equation (6.5) below) indicates that the diagonal (brace) elongation is

$$\delta_b = \frac{2\left[1 + \dfrac{a}{b}\right]\delta_1}{\left[1 + \left\{\dfrac{a+b}{2h}\right\}^2\right]^{0.5}} \tag{6.5}$$

which here is

$$\delta_b = \frac{2\left[1 + \dfrac{3.2}{2.8}\right] \times 3.6}{\left[1 + \left\{\dfrac{3.2 + 2.8}{2 \times 1.5}\right\}^2\right]^{0.5}} = 6.9 \text{ m}^2/\text{MN}.$$

Reference [2] equation (27) (equation (6.6) below) defines a dimensionless stiffness for bracing of

$$q = \frac{E_b A_b}{L_b} \frac{\delta_b^2}{l\delta_1}. \tag{6.6}$$

Here

$$q = \frac{200\,000 \times 0.002}{3.3} \times \frac{6.9^2}{7.5 \times 3.6} = 210.$$

Figure 6.4(b) indicates that for $\beta l = 0.45$ and $q = 210$ the value of w is about 0.024. Hence from equation (6.4)

$$W/P = 0.1 \times 0.024 = 0.0024 \text{ m/MN}$$

which is only 4% of the value without bracing.

Wright *et al.* [2] provides further formulae and charts for the derivation of some moments and stresses due to distortion and warping around the section.

6.4 SPACE FRAME ANALYSIS OF BOX-GIRDER

The distortion of a box-girder deck can be analysed without complexity by various finite element and space frame computer analyses. Figure

142   Box-girder decks

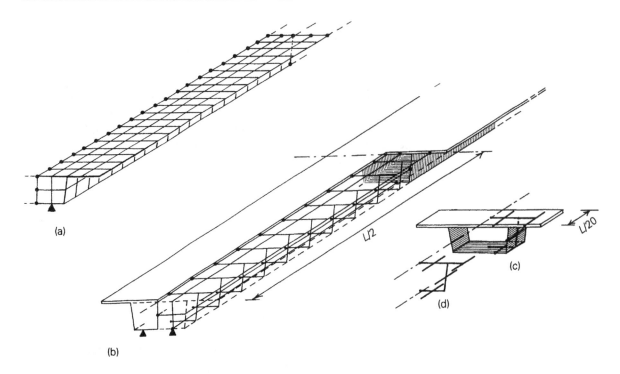

Fig. 6.6 Computer models of box girder: (a) finite element; (b)–(d) space frame.

6.6(a) and (b) illustrate finite element and space frame computer models of the deck in Fig. 6.3. The principles of the cruciform space frame method are explained in Section 7.3, and Section 7.3.1 demonstrates the derivation of member properties for the box girder of Fig. 6.3. The model in Fig. 6.6(b) represents half the width of the box-girder over half its length. The model has supports along the centre lines of the flanges which reproduce antisymmetry and it has supports on the transverse centre lines at midspan to reproduce symmetry. Figure 6.6(c) shows a short length of the box girder, and (d) illustrates the space frame members that reproduce the force system within the structure of (c). By keeping the model small it is possible to complete the modelling process in about one or two hours. However, it may be quicker in the long term to model the whole box-girder, so that loading and output do not have to be interpreted in terms of symmetric and antisymmetric components. The design load can then be placed in any position on the whole box model and the total force at any point in the box can be read directly from the computer output for the appropriate member. The space frame model can be built up in stages, if desired, in order to simulate skew, taper, curvature and so on.

Figure 6.5(b) illustrates the deformations calculated for the space frame when it is subjected to the antisymmetric loading from the 1 MN vehicle. It is evident that the distortion of the box is similar to that in Fig. 6.5(a) calculated with the BEF method. The deflection from the space frame is 0.03 m, which is mainly due to distortion with a small contribution from torsion.

Figure 6.5(c) illustrates the transverse bending moments calculated by the transverse space frame members. Figure 6.5(d) shows the longitudinal in-plane stresses at midspan derived from the axial forces and the in-plane bending moments in the longitudinal members of the space frame. All other forces in the box-girder due to distortion can also be read directly from the space frame output.

The numbers in brackets in Fig. 6.5(b), (c) and (d) are the output from a finite element analysis using the computer model in Fig. 6.6(a). The transverse bending stresses and longitudinal in-plane stresses from the finite elements are very close to the space frame results. The deflections in (b) from the finite elements are slightly smaller than from the space frame.

The significance of distortion on any bridge needs to be viewed within the context of the overall loading and behaviour of the structure. Figure 6.7 illustrates the results from the space frame analysis due to the total loading of dead load of 80 kN/m and distributed live load of 40 kN/m along one side. In Fig. 6.7(a) the average deflection is caused by the total load while the distortion is caused by the antisymmetric component of the live load. Figure 6.7(b) shows the transverse bending moments at midspan which are mainly caused by the local bending of the deck slab under the self weight of the slab and live load. Figure 6.7(c) shows the longitudinal bending stresses at midspan (the concrete

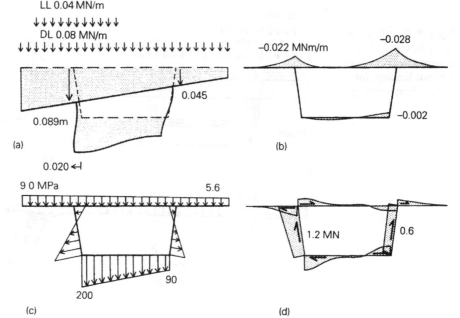

Fig. 6.7 Results from space frame analysis of composite steel box girder without bracing under direct load and eccentric live load: (a) deflections; (b) transverse bending moments; (c) longitudinal stresses; and (d) shear flows near support.

144 Box-girder decks

Fig. 6.8 Results from space frame analysis of composite steel box girder with cross-bracing, for comparison with Fig. 6.7.

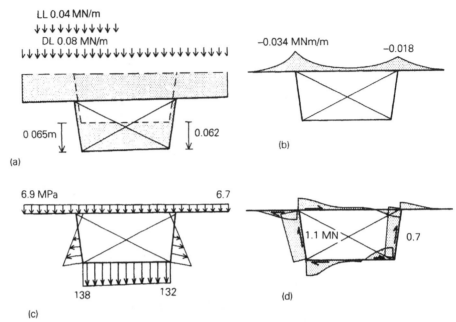

stresses are plotted to a scale adjusted by the modular ratio). Figure 6.7(d) shows the shear flows in the webs and flanges near the support.

The load distribution behaviour of the structure is greatly improved if some transverse bracing or framing is introduced at positions along the box. Cross-bracing can be introduced into the space frame model without difficulty. Figure 6.8 shows the results calculated by the space

Fig. 6.9 Results from space frame analysis of concrete box-girder without bracing, for comparison with Fig. 6.7.

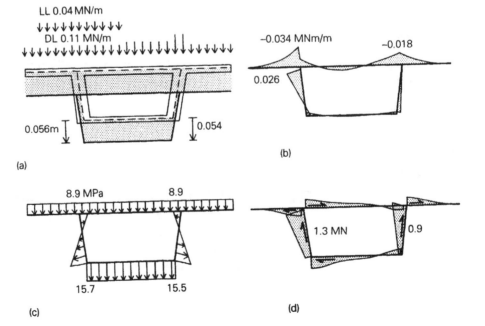

frame under the same loading as Fig. 6.7 when cross-bracing is included at 7.5 m spacing with area $A = 0.002\,\text{m}^2$. It is evident that this relatively small quantity of bracing has prevented most of the distortion, so that there is little variation in deflection across the box in Fig. 6.8(a), and little variation in longitudinal bending stresses in Fig. 6.8(b).

Prestressed concrete box-girders are relatively stiff against distortion, particularly if the webs are thick to accommodate prestressing tendons. Figure 6.9 illustrates the results calculated for a concrete box-girder of similar shape to the steel box-girder of Fig. 6.6, with 0.4 m thick webs and 0.15 m thick bottom flange. The dead load is greater due to the concrete structure, but the eccentric live load is as shown in Fig. 6.7. It is evident from the deflections in (a) and longitudinal stresses in (c) that distortion has had little effect.

6.5 GRILLAGE ANALYSIS OF BOX-GIRDER

Grillage analyses can be used for the global analyses of multiple box-girders and multispan decks, while a finite element analysis, space frame analysis, or BEF is used for the local analysis. The grillage has the advantage that it is very much faster for processing load cases and identifying the critical load case. The following paragraphs illustrate the derivation of section properties for one type of grillage and compare the results of an analysis for a single-span box-girder with the results from the space frame in Section 6.4. The comparison is reassuring and provides confidence for the use of the same grillage details for a multiple box-girder deck in Section 6.6. Several other grillage simulations may also work and a different method is demonstrated in Section 6.7. None provides a rigorous solution and the use of each is only justified by the closeness of its results to more rigorous methods. For each bridge attention has to be paid to the relevance of the approximations used. Particular care has to be taken if the bridge is curved, when distortional loading arises from dead load as well as live load.

Figure 6.10 illustrates a grillage for the box-girder of Fig. 6.3. The box-girder and slab have been modelled separately. The box-girder is represented by the longitudinal spine beam 1 with stiff outriggers 2. The stiffnesses of the box-girder in bending, torsion and distortion are represented only by the spine beam. The outriggers are used to transfer loads between the slab and the spine beam. The outriggers have moment releases at their outer ends so that they do not stiffen the slab. The slab is represented by the transverse members 3 which pass over the spine beam, and by the longitudinal members 4. The slab members are supported by the joints at the ends of the outriggers. In this grillage the twist rotation of the box longitudinal members 1 represents the average slope across the box; i.e. the relative vertical movement of the webs.

146 Box-girder decks

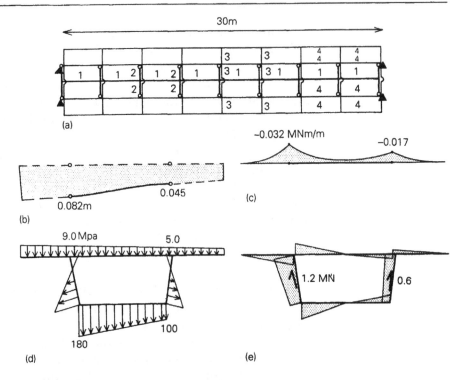

Fig. 6.10 Grillage analysis of composite steel box-girder: (a) grillage; and (b) to (e) results for comparison with Fig. 6.7.

The slopes of slab transverse members 3 reflect the transverse slope of the deck slab.

The distortion behaviour of the box-girder can be simulated by giving the spine beam members softened torsion constants. In Fig. 6.3(f) the distortion deflection W can be thought of as a rotation ϕ about the centre line where

$$\phi = \frac{2W}{a}. \qquad (6.7)$$

Hence, when the deck is loaded at midspan by concentrated antisymmetric loads, as in Fig. 6.3(e), the average twist over the length $l/2$ to each side is $(d\phi/dx) = (4W/al)$. The loads P subject the box to a torque Pa which is shared $Pa/2$ along the span to each side. Hence using equation (2.14) the equivalent torsion constant of the grillage longitudinal member is (ignoring the sign convention):

$$C = \frac{a^2 l}{8G} \frac{P}{W}. \qquad (6.8)$$

If $G = 0.4 E$ this can be written approximately as

$$C = \frac{a^2 l}{3.2E} \frac{P}{W}. \qquad (6.9)$$

The ratio (P/W) in the calculation of C can be obtained either from the space frame or finite element analysis or from equations (6.4) (equation (35) of the BEF method in reference [2]) which is given above. The results of the BEF analysis in Section 6.3 are demonstrated first, and then the results of the space frame method from Section 6.4. The analysis then proceeds with the space frame values because back-analysis of forces in details in the box-girder is more comprehensive from the space frame.

BEF equation (6.4) can be combined with equation (6.9) to obtain an equivalent torsion constant for distortion

$$C_d = \frac{a^2 l}{3.2E} \frac{8EI_b \beta^3}{w} = \frac{a^2 l I_c \beta^3}{1.6w} = \frac{a^2}{l^2} \frac{I_c(\beta l)^3}{1.6w}. \qquad (6.10)$$

Hence, for the example in Section 6.3, without cross-bracing, with $\beta = 0.06 \, \text{m}^{-1}$ and $w = 0.6$ (from Fig. 6.4(a)) we have

$$C_d = \frac{3.2^2 \times 30 \times 0.11 \times 0.06^3}{1.6 \times 0.6} = 0.0076 \, \text{m}^4.$$

This value can be compared to the pure torsion constant given by equation (2.20)

$$C_t = \frac{4A^2}{\oint \frac{ds}{t}} = \frac{4 \times 3^2 \times 1.5^2}{\frac{3.2}{0.03} + \frac{2.8}{0.02} + \frac{2 \times 1.5}{0.01}} = 0.15 \, \text{m}^4.$$

which shows that distortion has reduced the stiffness to about 1/20 of the true torsion stiffness.

The stiffness in distortion can also be compared to the torsional stiffness provided by the bending of two I-beams distance a apart subjected to antisymmetric loads P. Each deflects at midspan by $W = Pl^3/48EI$. If each has moment of inertia equal to half that of the box girder, $I = I_c/2$, the apparent torsion constant given by equation (6.9) is

$$C = \frac{a^2 l}{3.2E} \frac{P}{W} = \frac{a^2 l}{3.2E} \frac{48EI_c}{2l^3} = 7.5 \frac{a^2}{l^2} I_c$$

which in this case gives

$$C = 7.5 \times \frac{3.2^2}{30^2} \times 0.11 = 0.009 \, \text{m}^4.$$

Thus distortion of this box-girder, when unbraced, leads to an effective torsional stiffness which is less than that of two I-beams of the same total bending inertia. Thus if this box-girder without bracing were to be modelled with a grillage with two longitudinal members along the webs (each of inertia $I_c/2$), the antisymmetric flexure of these members would provide too much resistance to distortion. However when a box-girder has internal bracing, the distortion is much less significant and a grillage can be used with two longitudinal members along the webs, as is illustrated in Section 6.7.

If one wanted to include the torsion rotations (which are ignored in the BEF method) as well as distortion the combined distortion + torsion constant C_{dt} is given by

$$\frac{1}{C_{dt}} = \frac{1}{C_d} + \frac{1}{C_t} \qquad (6.11)$$

which here gives

$$C_{dt} = \frac{1}{\frac{1}{0.0076} + \frac{1}{0.15}} = 0.0072 \text{ m}^4.$$

When cross-bracing is included the torsion component becomes more dominant. Equation (6.10) with $l = 7.5$ m and $w = 0.024$ leads to $C_d = 0.091$ m^4. When the torsion constant $C_t = 0.15$ m^4 is included, equation (6.11) gives $C_{dt} = 0.05$ m^4.

This demonstration now reverts to the results of the space frame analysis in Section 6.4. The displacement W output from the space frame analysis includes distortion and torsion displacements so that equation (6.8) can be used to obtain C_{dt}. In the example of Section 6.4, illustrated in Fig. 6.5, $W = 0.030$ m when $P = 0.5$ MN; hence

$$C_{dt} = \frac{3.2^2 \times 30 \times 0.5}{8 \times 80\,000 \times 0.030} = 0.0080 \text{ m}^4.$$

The bending inertia of spine beam members 1 in Fig. 6.10 is equal to that of the whole box. Hence

$$I_1 = 0.11 \text{ m}^4$$

and its shear area is equal to the area of the two webs

$$A_{S1} = 1.5 \times 0.010 \times 2 = 0.03 \text{ m}^2.$$

The outrigger members are given the same stiffnesses as the spine beam.

Member types 3 and 4 in Fig. 6.10 representing slab transversely and longitudinally have bending and torsion stiffnesses based on equations (3.14) and (3.15). But if the slab has been included in the space frame derivation of C_{dt} and in I, the longitudinal slab members should be given only nominal stiffnesses. In this example nominal stiffnesses have been used for the longitudinal slab members, and the members only assist load descriptions. When the box-girder forms part of a multiple box-girder deck, as described in Section 6.6, the stiffnesses of transverse slab members have a direct influence on how loads are transferred between spine beams.

Figure 6.10 illustrates some of the calculated effects from this grillage

Fig. 6.11 Results from grillage analysis of composite steel box girder with bracing, for comparison with Fig. 6.8.

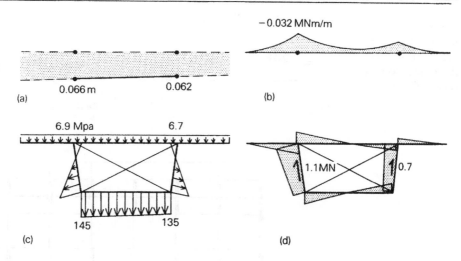

when it is subjected to the same loading as shown in Fig. 6.7 with dead load of 80 kN/m and eccentric live load of 40 kN/m. Figure 6.10(b) shows the vertical deflections of the webs. Figure 6.10(c) illustrates the transverse bending moments in the slab excluding distortion moments from space frame. Figure 6.10(d) illustrates the longitudinal stresses. The stresses at the centres of the flanges were calculated from the bending moment in the spine beam and the section properties of the box girder. The variations in stresses across the widths of the flanges were derived by scaling the centre line stresses in proportion to the deflections in Fig. 6.10(b). The shear flows in Fig. 6.10(e) were calculated from the shear force and torsion in the spine beam using simple beam theory, as described in Chapter 2.

The results in Fig. 6.10 from the grillage can be compared with those in Fig. 6.7 from the space frame analysis. It is evident that the grillage results compare well with the space frame.

If the box-girder is braced the same procedures can be followed. The torsion constant is again based on the space frame analysis (or BEF) using equation (6.9). In this case $W = 0.0024$ m when $P = 0.5$ MN and $C_{dt} = 0.10$ m^4. The results from the grillage analysis of the braced box-girder are illustrated in Fig. 6.11. It can be seen that these are remarkably similar to those in Fig. 6.8 from the space frame analysis. Transverse bending stresses can be derived from a plane frame analysis of the cross-section, as described in Section 6.6, or from the space frame.

6.6 GRILLAGE ANALYSIS OF MULTIPLE BOX-GIRDER DECK

A bridge deck constructed with a number of box-girders can be analysed with a grillage based on the same principles as the preceding grillage for

Fig. 6.12 Grillage for multiple box-girder deck. (a) Deck section; (b) grillage section; (c) grillage mesh.

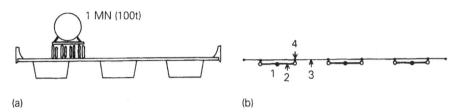

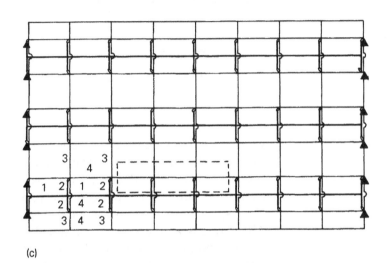

the single box-girder. Figure 6.12 illustrates a multiple box-girder deck and a possible grillage model. Each box-girder is represented by a longitudinal spine beam 1 with outriggers 2, (which have moment releases at outside ends). The slab is represented by transverse members 3, which pass over the spine beams and are supported at the ends of the outriggers. The longitudinal members 4 are included mainly to assist load description. The section properties are derived in Section 6.5.

Figure 6.13 illustrates the results from the grillage for the multiple box beam deck when a 1 MN (100 tonne) vehicle stands in the position shown in Fig. 6.12. The deflections at midspan are shown in Fig. 6.13(a), while Fig. 6.13(b) shows the transverse bending stresses in the slab (excluding distortion moments from space frame and local wheel effects) and Fig. 6.13(c) shows longitudinal bending stresses.

The grillage may be used to determine the critical load cases for each box-girder. The local stresses in each box-girder are then found from a space frame analysis under the same system of forces.

Fig. 6.13 Results from grillage of multiple box-girder deck: (a) deflections; (b) transverse bending moments; and (c) bending stresses.

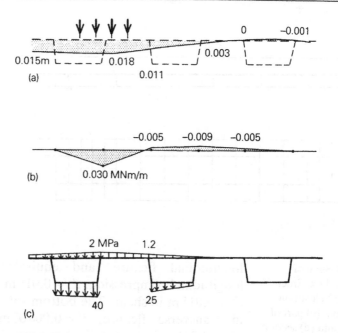

6.7 GRILLAGE ANALYSIS OF A MULTISPAN BOX-GIRDER

A box-girder which is relatively stiff against distortion can be analysed with a shear-flexible grillage which represents the box with two longitudinal members, as shown in Fig. 6.14. In this grillage the twist rotation of each of the main longitudinals represents the rotation of each web as a whole; i.e. the sway of the top of the web relative to the bottom as shown in Fig. 6.15(a). This rotation cannot also reproduce the rotation of the top slab at the top of the web. The transverse members are given bending inertias and shear flexibility, based on Chapter 5, which link the rotations of the two webs during distortion. The torsion constants are calculated below as for a non-distorting section, and the results are compared with analyses by space frame and BEF. In the example of Fig. 6.14 the box is stiffened by frames at 4 m spacing and bracing at 20 m spacing. In the grillage transverse members are placed at 10 m centres (closer near supports).

The flanges and webs of the box-girder in Fig. 6.14 have the following equivalent properties, using Section 5.4.5 and its notation and taking account of stiffening. The top flange has equivalent inertia $i' = 0.0008 \text{ m}^4/\text{m}$ in transverse flexure, $i' = 0.00008 \text{ m}^4/\text{m}$ in

152 Box-girder decks

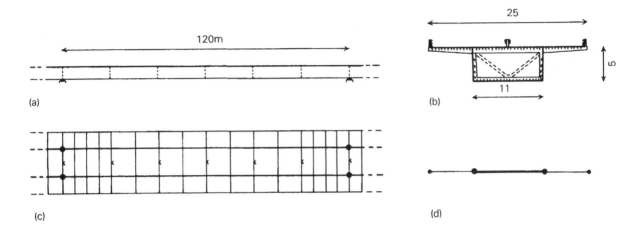

Fig. 6.14 Span of multispan box-girder bridge: (a) elevation; (b) section; (c) part of grillage; and (d) section of grillage.

longitudinal flexure, and equivalent thickness $d' = 0.022$ m in longitudinal compression, $d' = 0.018$ m in transverse compression, and $d' = 0.015$ m in shear. The bottom stiffened plate has $i'' = 0.0002\,\text{m}^4/\text{m}$ in transverse flexure, $i'' = 0.00008\,\text{m}^4/\text{m}$ in longitudinal flexure, $d'' = 0.021$ m in longutidinal compression, $d'' = 0.017$ m in transverse compression, and $d'' = 0.014$ m in shear. The webs are assumed to have $i_w = 0.0003\,\text{m}^4/\text{m}$ in transverse flexure, $i_w = 0.0008\,\text{m}^4/\text{m}$ in longitudinal flexure, $d_w = 0.020$ m in longitudinal compression, $d_w = 0.016$ m in transverse compression, and $d_w = 0.013$ m in shear. The bracing at 20 m spacing has a cross-sectional area of $0.005\,\text{m}^2$.

The main longitudinal members of the grillage have inertias based on the effective thicknesses of the stiffened plates in longitudinal compression, whence

$$I_1 = I_2 = 2.3\,\text{m}^4.$$

The transverse members at 10 m centres have bending inertias based on equation (5.4). Hence

$$I_t = \frac{5^2 \times 0.018 \times 0.017}{(0.018 + 0.017)} \times 10 = 2.4\,\text{m}^4.$$

The shear area of transverse members is obtained with a plane frame analysis as shown in Fig. 6.15(a). The frame members have inertias based on i', i'' and i_w in transverse flexure. It is found that when the

Fig. 6.15 Plane frame of cross-section of deck of Fig. 6.14: (a) for derivation of shear areas and (b) for local moments.

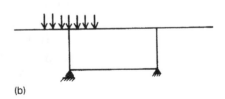

frame represents 10 m of box without bracing a shear force of $S = 8.6$ MN is required to cause a unit shear distortion $w_s = 1$ m. Hence, using equation (5.7)

$$A_S = \frac{Sl}{Gw_s} = \frac{8.6 \times 11}{80\,000 \times 1} = 0.0012 \, \text{m}^2.$$

When bracing is included the frame members have axial areas corresponding to the amount of frame and plate which interacts with the bracing after taking account of shear lag, which is explained in Chapter 8. With bracing of area $0.005 \, \text{m}^2$ in the plane frame it is found that a shear force of $S = 33$ MN is required to cause unit shear distortion. Hence

$$A_S = \frac{33 \times 11}{80\,000 \times 1} = 0.0045 \, \text{m}^2.$$

The torsion constant of a non-distorting box is given by equation (2.20), which here gives

$$C_t = \frac{4 \times 11^2 \times 5^2}{\left(\dfrac{11}{0.015} + \dfrac{11}{0.014} + \dfrac{2 \times 5}{0.013}\right)} = 5.2 \, \text{m}^4.$$

In the grillage the torsion stiffness is shared between longitudinal members and transverse members. If it is assumed to be shared equally between longitudinal and transverse members the two longitudinal members together have $2.6 \, \text{m}^4$. Hence each has

$$C_1 = C_2 = 1.3 \, \text{m}^4.$$

The longitudinal members are 11 m apart, so that the torsion constant per unit width is

$$c = \frac{2.6}{11} \, \text{m}^4/\text{m}$$

and transverse members at 10 m spacing are given torsion constants of

$$C = 10 \times \frac{2.6}{11} = 2.4 \, \text{m}^4.$$

The above values of torsion constant are smaller than values derived by equation (5.5) because equation (2.2) includes shear of the webs while equation (5.5) does not. In a multicellular deck the difference is usually negligible. Here the difference also has little influence on the distribution of forces because little torsion deformation occurs.

Some results from the multispan grillage are shown in Fig. 6.16. Figure 6.16(a) is the bending moment diagram for the two main

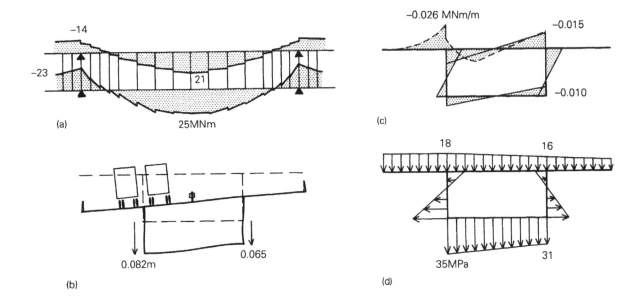

Fig. 6.16 Results from grillage of Fig. 6.14: (a) moments along longitudinal members; (b) deflections; (c) transverse moments; and (d) longitudinal bending stresses.

longitudinal members when the span is loaded on one side with a distributed load of 45 kN/m. Figure 6.16(b) shows the deflection at a section 10 m from midspan where there is no bracing. Figure 6.16(c) shows the longitudinal bending stresses. The transverse moments in Fig. 6.16(c) were derived with the frame in Fig. 6.15, by subjecting the frame as in Fig. 6.15(a) to the shear force in the transverse grillage member, and adding the local moments from the frame as shown in Fig. 6.15(b). A similar back-analysis of the transverse member at midspan using the plane frame of Fig. 6.15 with bracing provides an indication of the forces in the bracing.

The behaviour of this grillage simulation has been compared to a space frame analysis and the BEF method for the deck over a simply supported span of 90 m, to correspond with the distance between points of contraflexure in Fig. 6.16(a). The space frame had a similar model to Fig. 6.6, but with dimensions and section properties calculated for this structure. Under the action of antisymmetric loads of 0.5 MN at midspan the 90 m grillage deflected on each side by 0.003 m while the space frame deflected by 0.004 m. Using equation (6.8) it is found that the grillage and space frame have an equivalent distortion + torsion constant $C_{dt} = 3.0 \, \text{m}^4$ and $2.2 \, \text{m}^4$, respectively. If the BEF method is used with the bracing bay length of 20 m it is found that the dimensionless parameter $\beta l = 1.0$ and $q = 0.5$, and equation (6.10) gives $C_d = 2.0 \, \text{m}^4$. If this is combined with $C_t = 5.2 \, \text{m}^4$, calculated above, using equation (6.11) we obtain $C_{dt} = 1.5 \, \text{m}^4$. This range of values appears to be large. However, all these values of C_{dt} are indicative of a stiff structure.

Grillage analysis of a multispan box-girder

Fig. 6.17 Multiple steel box-girder viaduct at Kings Langley, England; designed by Tony Gee and Partners. Photograph E. C. Hambly.

REFERENCES

1. Nakai, H. and Yoo, C.H. (1988) *Analysis and Design of Curved Steel Bridges*, McGraw-Hill, New York.
2. Wright, R.N., Abdel-Samad, S.R. and Robinson, A.R. (1968) BEF for analysis of box girders, *Journal of the Structural Division of the ASCE*, **94**, ST7, 1719–43.
3. Maisel, B.I. and Roll, F. (1974) 'Methods of analysis and design of concrete box beams with side cantilevers', Technical Report 42.494, Cement and Concrete Association, UK.
4. Lee, D.J. and Richmond, B. (1988) 'Bridges', in *Civil Engineer's Reference Book*, (ed. L.S. Blake) Butterworths, Guildford.
5. Post-Tensioning Institute and Prestressed Concrete Institute (1978) *Precast Segmented Box Girder Bridge Manual*, PCI, Chicago.
6. British Standard BS5400: Part 3: 1982, 'Steel, concrete and composite bridges, Part 3 Code of Practice for Design of Steel Bridges', British Standards Institution.

7   Space frame methods and slab membrane action

7.1 TRUSS SPACE FRAME

An engineer can choose from a wide variety of possible computer models for the design or assessment of a bridge. The precise choice depends on which strengths of the structure he wishes to utilize, and on how local loads and analyses are related to the global analyses.

The three-dimensional behaviour of some bridges may on occasions be analysed more straightforwardly with a three-dimensional space frame than with an equivalent two-dimensional grillage. The preparation of a space frame model is generally more complicated and time consuming than a grillage but the geometric similarity of the space frame to the three-dimensional structure can make it easier to interpret. As computing facilities improve it becomes easier to use more complicated analyses. However it is still essential to keep the computer model as simple as is practicable, otherwise the engineer may find that most of his time is absorbed in unscrambling his computer results rather than in design.

Figure 7.1(a) illustrates a relatively simple space frame idealization for a straight composite steel girder bridge. Figure 7.1(b) and (c) shows half sections of the deck and of the space frame. In Fig. 7.1 the primary structural members are shown by continuous lines while nominal members for load application are shown dashed. The reinforced concrete slab has been idealized by longitudinal member A, transverse member B and one diagonal C in each bay for transferring shear. The girder has been represented by longitudinal members D for the top half (coincident with A) and E for the bottom half, with vertical member F and diagonal G for transferring shear. Space frame members need to be located at cross-frames, diaphragms and dominant stiffeners. The diagonals G on the girder are shown in this example in the direction of tension under dead load in order to simulate the tension field in the web. It is assumed in this case that the stiffness in the compression diagonal direction can be ignored since the diagonal may buckle at the

158 Space frame methods and slab membrane action

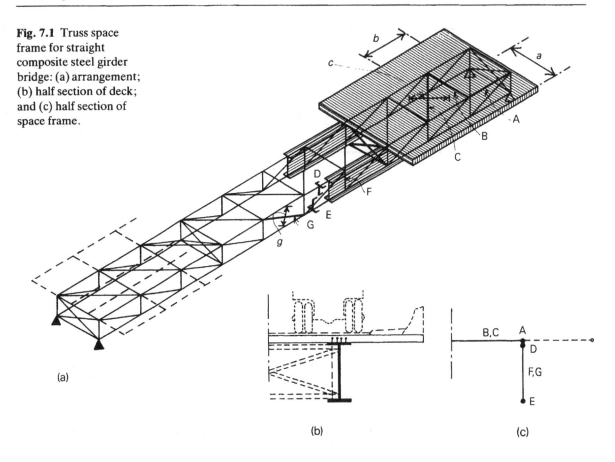

Fig. 7.1 Truss space frame for straight composite steel girder bridge: (a) arrangement; (b) half section of deck; and (c) half section of space frame.

ultimate limit state. In the concrete slab the diagonal C is placed in the compression direction, on the basis that the reinforced concrete will be stiffer in compression than when cracked in the tension direction.

In the example of Fig. 7.1 the members have section properties calculated for:

- A: slab over half width dimensioned a
- B: slab over length b between bay centre lines
- C: slab over diagonal width c
- D: top half of girder
- E: bottom half of girder
- F: web plate over width $f (= b)$
- G: web plate over inclined width g

The stresses in the structure can be calculated directly from the forces and moments output by the program. The shear stress in the web can be calculated from the vertical component of the forces in diagonal G.

The straight deck of Fig. 7.1 can probably be analysed more easily as a grillage, rather than as a space frame, but if the deck is curved, as in

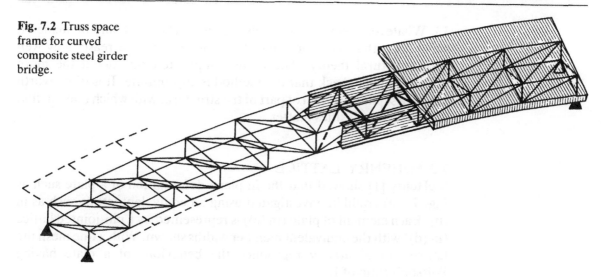

Fig. 7.2 Truss space frame for curved composite steel girder bridge.

Fig. 7.2, it may be easier to design the bracing between beams, by modelling them in a space frame. Steel girders have very low stiffness against torsion, and the curved structure has to have bracing between the girders to make the structure stiff against torsional loads.

It is possible to use more complicated space frames with members in both diagonal directions, as shown in Fig. 7.3(a). Each rectangle of members with diagonals then represents a rectangle of plate, as is explained in Section 7.2 for a McHenry lattice. However the sizing of the members is much more complicated because the diagonals carry some of the longitudinal loads. It is likely to be easier to model such a deck with a cruciform space frame, as shown in Fig. 7.3(b) and described in Section 7.3, or with finite elements, as described in Chapter

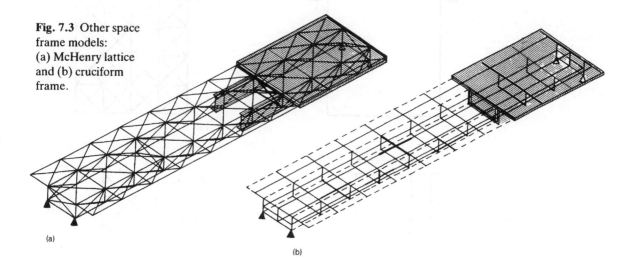

Fig. 7.3 Other space frame models: (a) McHenry lattice and (b) cruciform frame.

13. Whatever method of modelling is used, checks should be made on small parts that the properties of the model compare reasonably well with structural theory. The more complicated the model the more difficult it is to check that the method is appropriate. It is often worth having a separate model of part of the structure, with which one can find out what complexities matter.

7.2 McHENRY LATTICE

McHenry [1] showed that the in-plane deformation of a plate such as Fig. 7.4(a) could be investigated using an equivalent lattice such as in (b). Each element of plate (in (c)) is represented by a pin-jointed lattice (in (d)) with the equivalent member widths shown. With a fine mesh the lattice can accurately reproduce the behaviour of a plate having Poisson's ratio of 1/3.

A McHenry lattice can be incorporated into a space frame to produce a three-dimensional model of a bridge. However if the grid mesh is not close to square the lattice does not behave like a piece of plate. It is then likely to be easier and more accurate to use a cruciform space frame.

Fig. 7.4 McHenry lattice.

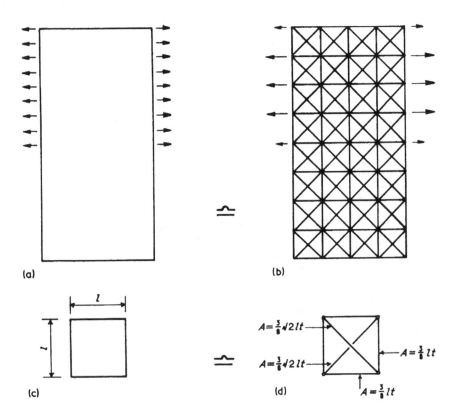

Fig. 7.5 Cruciform element.

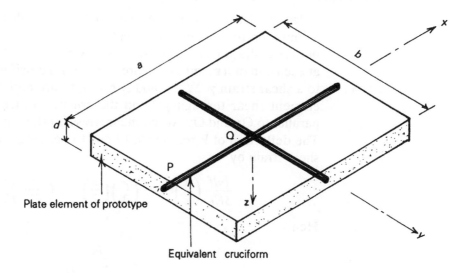

7.3 CRUCIFORM SPACE FRAME

7.3.1 Stiffnesses

Beam-and-slab and cellular bridge decks can be modelled in three dimensions with a space frame which has its members arranged in cruciforms, as illustrated in Fig. 7.3(b). The slabs and webs of the prototype are notionally subdivided into rectangular elements (of as nearly the same size as possible), and each of these is represented in the space frame model by a cruciform of beam members as shown in Fig. 7.5. Since there is no interaction in the cruciform between compression and bending in one direction with the compression and bending in the other direction, the model effectively assumes that Poisson's ratio is zero (except where used in the shear modulus).

The section properties attributed to the cruciform members depend on whether shear flexibility is considered by the space frame, or not. Equations are derived below, first for shear-rigid members and then for shear-flexible members.

Fig. 7.6 Shear of cruciform element.

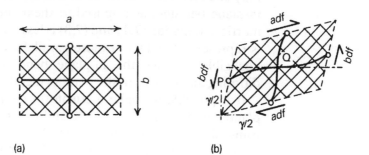

The in-plane stiffnesses of the cruciform members are based on the assumption that under shear loading there are points of contraflexure between adjacent cruciforms, i.e. midway between the cruciform joints at each end of a member. Figure 7.6 shows a cruciform being subjected to a shear strain γ. When bodily rotations are removed all sides of the element shear through $\gamma/2$ and the cruciform members at Q remain parallel to Ox and Oy. If the shear stress is f the shear force at P is bdf. The deflection of P relative to Q due to bending can be related to the shear strain by

$$\frac{bdf}{3EI_z}\left(\frac{a}{2}\right)^3 = \left(\frac{\gamma}{2}\right)\left(\frac{a}{2}\right) = \left(\frac{f}{2G}\right)\left(\frac{a}{2}\right).$$

Hence

$$I_z = \frac{ba^2 d}{6(E/G)}. \tag{7.1}$$

If Poisson's ratio $\nu = 0.25$ the ratio $(E/G) = 2.5$ and

$$I_z = \frac{ba^2 d}{15}. \tag{7.2}$$

The four section properties of members PQ in Fig. 7.5 are then

$$\left.\begin{array}{ll}
\text{Compression area} & A_x = bd \\[4pt]
\text{Torsion constant} & C = \dfrac{bd^3}{6} \\[6pt]
\text{Out-of-plane bending inertia} & I_y = \dfrac{bd^3}{12} \\[6pt]
\text{In-plane bending inertia} & I_z = \dfrac{ba^2 d}{15} = \left(\dfrac{a}{b}\right)^2 \dfrac{b^3 d}{15}
\end{array}\right\} \tag{7.3}$$

When output forces are converted back into stresses the normal section properties are used including section modulus $Z_z = b^2 d/6$.

If a very coarse frame is used so that a piece of plate is represented by only one member, that member will be over-flexible under the action of in-plane bending, as opposed to shear, because the in-plane bending inertia of equation (7.1) and (7.3) is less than $b^3 d/12$. If the piece of plate is represented by two or more parallel members this over-flexibility is negligible because the in-plane bending is carried mainly by the differing axial forces in the parallel members.

It is possible to make a cruciform member reproduce both the in-plane bending with $I_z = b^3 d/12$ and the in-plane shear deformation by using shear-flexibility with a modified shear area. In this case the deflection of point P in Fig. 7.6(b) relative to Q due to bending and shear can be related to shear strain by

$$\frac{bdf}{3EI_z}\left(\frac{a}{2}\right)^3 + \frac{bdf}{A_y G}\left(\frac{a}{2}\right) = \left(\frac{\gamma}{2}\right)\left(\frac{a}{2}\right) = \left(\frac{f}{2G}\right)\left(\frac{a}{2}\right).$$

Putting $I_z = b^3 d/12$ and rearranging we obtain

$$A_y = \frac{bd}{(0.5 - (G/E)(a/b)^2)}. \tag{7.4}$$

If Poisson's ratio $\nu = 0.25$ with $(E/G) = 2.5$

$$A_y = \frac{bd}{(0.5 - 0.4(a/b)^2)}. \tag{7.5}$$

The six section properties of member PQ in Fig. 7.5 are now

$$\left.\begin{array}{ll} \text{Compression area} & A_x = bd \\[4pt] \text{In-plane shear area} & A_y = \dfrac{bd}{(0.5 - 0.4(a/b)^2)} \\[4pt] \text{Out-of-plane shear area} & A_z = bd \\[4pt] \text{Torsion constant} & C = \dfrac{bd^3}{6} \\[4pt] \text{Out-of-plane bending inertia} & I_y = \dfrac{bd^3}{12} \\[4pt] \text{In-plane bending inertia} & I_z = \dfrac{b^3 d}{12} \end{array}\right\} \tag{7.6}$$

It has been found that remarkably coarse meshes can be used without serious error. Figure 7.7(a) and (b) illustrate two models of half the length of a wide flange of a beam with flange width equal to 1/4 of the overall length. The flanges are being subjected to the in-plane shear

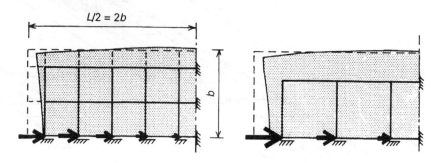

Fig. 7.7 Shear lag in cruciform models of wide flange.

from bending of the beam under a distributed load. The effective flange width ratios of models (a) and (b) are found to be 0.7, which compare well with values from Fig. 8.4 and reference [2].

It can be seen that the expression for shear area A_y in equation (7.4) is a hyperbolic function, and that it tends to infinity for $(a/b)^2$ close to $(0.5E/G)$ and is negative for $(a/b)^2$ greater than $(0.5E/G)$. Unless A_y can be input as rigid, this ratio of $(a/b)^2 = (0.5E/G)$ should be avoided. It is generally best to keep the mesh as close to square as possible. If negative values of A_y are used a check should be made that the computer program treats them as negative.

The cruciform model can also be used with members in all three directions to simulate a solid block of material as is illustrated in Chapter 14. In this case I_y as well as I_z in equation (7.3) has the modified inertia, or A_z as well as A_y in equation (7.6) has the modified shear area.

The author has used cruciform space frame models on many occasions, particularly when analyses of very complex structures have been required urgently. On each occasion he has calibrated his model at various stages by comparing the stiffnesses of parts with established structural theory. When complex analyses have later been compared with finite element methods the results have been found satisfactory.

7.3.2 Example of box-girder deck

Figure 7.8 illustrates the cruciform space frame model used to analyse the composite box-girder deck that is examined in Sections 6.3 and 6.4 and illustrated in Fig. 6.3.

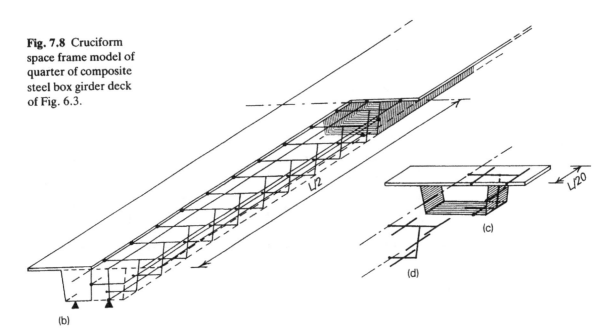

Fig. 7.8 Cruciform space frame model of quarter of composite steel box girder deck of Fig. 6.3.

The deck has a span of 30 m, and hence the half-span model in Fig. 7.8 is 15 m long. Transverse frames of members are located at spacings of 1.5 m along the deck. The bottom flange is 0.02 m thick and 2.8 m wide so that the half width in the model is 1.4 m. Hence for the longutidinal members in the bottom flange

$$a = 1.5 \text{ m} \quad b = 1.4 \text{ m} \quad d = 0.020 \text{ m}.$$

Treating members as shear-rigid, equation (7.3) gives

$$A_x = 1.4 \times 0.02 = 0.028 \text{ m}^2$$

$$C = \frac{1.4 \times 0.02^3}{6} = 1.9 \times 10^{-6} \text{ m}^4$$

$$I_y = \frac{1.4 \times 0.02^3}{12} = 0.9 \times 10^{-6} \text{ m}^4$$

$$I_z = \frac{1.4 \times 1.5^2 \times 0.02}{15} = 0.0042 \text{ m}^4.$$

Transverse members in the bottom flange have

$$a = 1.4 \text{ m} \quad b = 1.5 \text{ m} \quad d = 0.020 \text{ m}$$

so that equation (7.3) gives

$$A_x = 1.5 \times 0.02 = 0.030 \text{ m}^2$$

$$C = \frac{1.5 \times 0.02^3}{6} = 2.0 \times 10^{-6} \text{ m}^4$$

$$I_y = \frac{1.5 \times 0.02^3}{12} = 1.0 \times 10^{-6} \text{ m}^4$$

$$I_z = \frac{1.5 \times 1.4^2 \times 0.02}{15} = 0.0039 \text{ m}^4.$$

The web is 0.010 m thick and 1.5 m high, so that longitudinal and transverse (vertical) members all have $a = b = 1.5$ m and $d = 0.010$ m. The section properties are calculated as above, except that care has to be taken about the orientation of in-plane and out-of-plane axes. The longitudinal members have axes rotated through a small 'beta angle' due to the inclination of the web.

The slab is 0.2 m thick. Longitudinal members representing the slab above the box have width $b = 1.6$ m, while longitudinal members representing cantilever have $b = 1.7$ m.

7.4 SLAB MEMBRANE ACTION

It was shown in Section 4.10 that the conventional plane grillage does not reproduce transfer of in-plane shear across the slab strips between

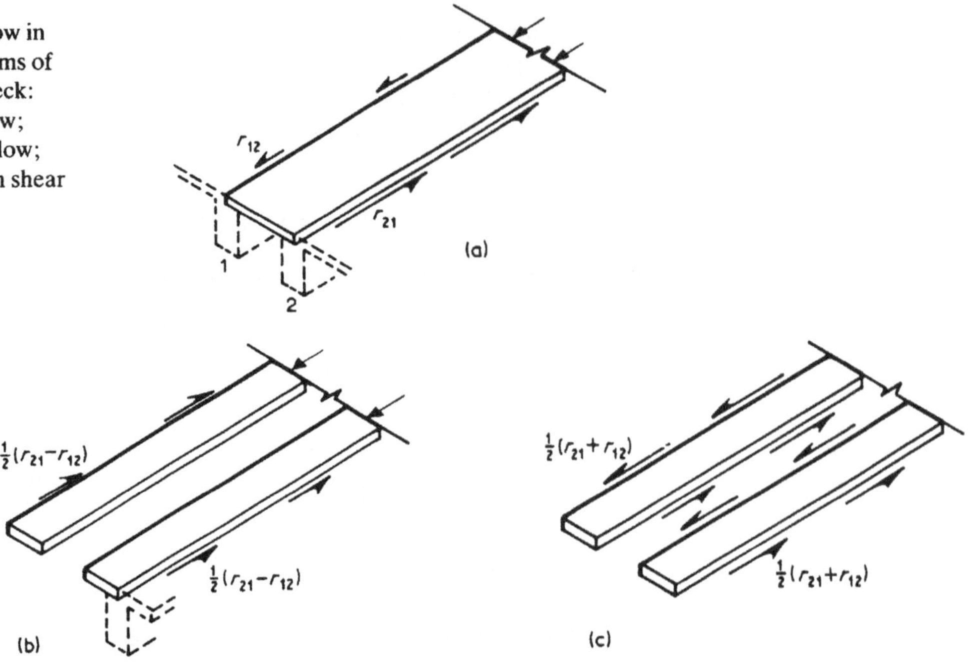

166   Space frame methods and slab membrane action

Fig. 7.9 Shear flow in slab between beams of beam-and-slab deck: (a) total shear flow; (b) flange shear flow; and (c) interbeam shear flow.

the longitudinal beams of a beam-and-slab deck. Figure 7.9(a) shows such a slab strip subjected to in-plane shear flows r_{12} and r_{21} along its edge. These shear flows can be split into the symmetric shear flows in (b) and antisymmetric shear flows in (c). The conventional grillage analysis simulates the shear flows of (b) by considering the two halves of the slab strip as flanges of the T-beams to each side. It is the antisymmetric interbeam shear flow of (c) that is ignored.

The antisymmetric shear flows of Fig. 7.9(c) can cause two types of in-plane deformation depending on whether they are accompanied by transverse compression forces along the edges. Figure 7.10(a) shows a narrow slab strip subjected to antisymmetric shear flows in the absence of transverse forces, and the strip deflects by in-plane bending with cross-sections fanning radially. In contrast, (b) shows a thin slab strip distorting in-plane by shear deformation with cross-sections remaining parallel and rectangular elements shearing to parallelograms. An isolated narrow strip is very much stiffer in shear than in bending, so that

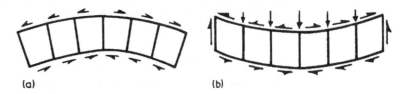

Fig. 7.10 In-plane deformation of strip of slab: (a) bending, and (b) shear.

Fig. 7.11 Sideways flexure of channel: (a) elevation; (b) section; and (c) plan of top slab.

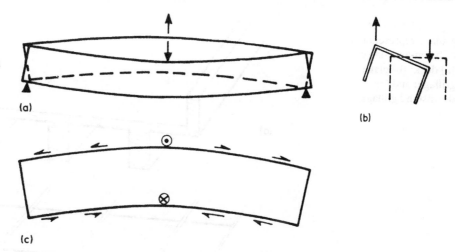

if the slab represents the back of a narrow channel under antisymmetric loads, as in Fig. 7.11, the edge shear flows cause it to deflect sideways in bending (in direction towards the stretched edge). However, if the strip is stiffened against sideways displacement by numerous other strips forming a wide slab, the edges remain virtually straight and the slab distorts in 'trapezoidal shear' as in Fig. 7.12. This is a combination of the in-plane bending of Fig. 7.10(a) and in-plane shear of Fig. 7.10(b). Because the strip has much greater stiffness for in-plane shear than for in-plane bending, the dominating forces and stresses associated with such distortion relate to shear.

7.5 DOWNSTAND GRILLAGE

7.5.1 Geometry and stiffnesses

Beam-and-slab bridges can be analysed as three-dimensional structures by a space frame analysis which is a simple extension of the grillage. Instead of a deck such as in Fig. 7.13(a) being represented by a plane

Fig. 7.12 'Trapezoidal shear' of strip of slab.

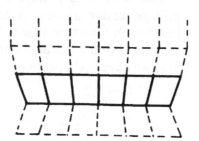

168 Space frame methods and slab membrane action

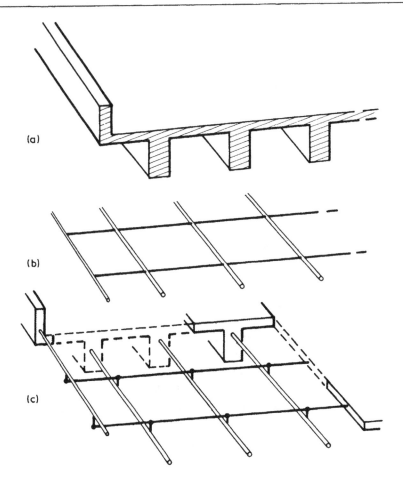

Fig. 7.13 Downstand grillage representation of beam-and-slab deck: (a) deck; (b) plane grillage; and (c) downstand grillage.

grillage in (b), a space frame is used, as in (c). The mesh of the space frame in plan is identical to the grillage, but the various transverse and longitudinal members are placed coincident with the line of the centroids of the downstand or upstand member they represent. For this reason the space frame is here referred to as a 'downstand grillage'. The longitudinal and transverse members are joined by vertical members which, being short, are very stiff in bending.

The downstand grillage behaves in a similar fashion to the plane grillage under actions of transverse and longitudinal torsion and bending in a vertical plane. Consequently, the section properties for these actions are calculated in the same way. Thus for the deck of Fig. 7.14 we have, as in Section 4.5.2,

$$I_x = 0.21 \qquad C_x = 0.0032$$
$$I_y = \frac{4 \times 0.2^3}{12} = 0.0027 \qquad C_y = \frac{4 \times 0.2^3}{6} = 0.0053.$$

Fig. 7.14 Dimensions of composite steel and concrete deck: (a) section of deck; (b) longitudinal member; and (c) transverse member.

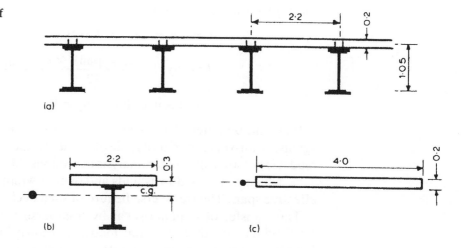

Vertical deflection of the downstand beams causes joints in the plane of the slab to move longitudinally as in Fig. 7.15(a). These warping displacements generate in-plane shear forces in the transverse members if the members are given the shear areas of the cross-section of slab they represent and if they are forced to shear as in Fig. 7.15(b). Curvilinear distortion of the mesh as in Fig. 7.15(c) must be prevented by restraining all slab joints against sideways displacement v and rotation θ_z about the vertical axis. In addition, transverse members are given very high inertias for in-plane bending so that shear deformation dominates. Thus

Fig. 7.15 In-plane shear of transverse members of downstand grillage representing slab: (a) warping displacement due to bending of longitudinal members; (b) plan view of in-plane shear of transverse members; and (c) erroneous in-plane flexure of transverse members.

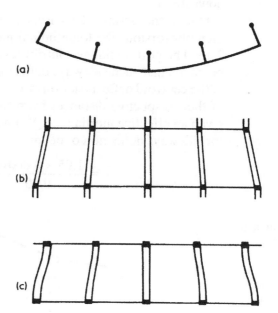

for in-plane shear and moment in the slab of Fig. 7.14,

$$A = bd = 4.0 \times 0.2 = 0.8$$

$$I = \text{say}, \frac{(\text{deck span})^3 \times d}{12} = 800$$

Joints in slab $v = \theta_z = 0$.

It should be noted that because the joints in the slab are restrained against transverse horizontal displacement, the downstand grillage model is not appropriate for analysis of sections, like the U-beam in Fig. 7.11, which has deep downstands and overall width less than about 1/3 effective span. The transverse flexure of such decks is significant.

The transfer of in-plane shear by transverse members subjects the longitudinal members to axial loads, as shown in Fig. 4.13(b). The longitudinal members must be able to stretch (or compress) under these axial forces and so each is given the cross-sectional area of the part of deck it represents, including slab flanges. For the deck of Fig. 7.14, with modular ratio $m = 7$ for steel beam

$$A_x = \text{longitudinal T-beam area} = 0.09 \times 7 + 2.2 \times 0.2 = 1.27.$$

If the longitudinal beams are spaced at more than 1/5 effective span, shear lag reduces the effective width of flange and interbeam shear transfer is small. The width of slab considered as flange to each side of the beam should be limited to about 1/10 of span, as explained in Chapter 8. Any slab between downstands and not included in their flanges is considered as a separate longitudinal beam without downstand.

Under the action of transverse member bending and longitudinal member torsion, the longitudinal members deflect sideways as in Fig. 7.16. The bottom flange contributes most of the flexural stiffness of the beam against such sideways deflection. Since the longitudinal member at the centroid deflects sideways less than the bottom flange in ratio z_c/z_b of their respective distances from the top slab, the member should be given an effective inertia $(z_b/z_c)^2$ times that of the bottom flange. Hence for sideways deflection of beam

$$I = \frac{1.05 \times 0.00068}{0.30^2} \times 7 = 0.058.$$

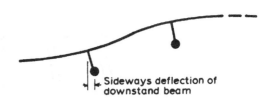

Fig. 7.16 Part section of downstand grillage showing sideways deflection of longitudinal beam due to transverse bending.

The vertical members behave in a similar manner to transverse slab members except that the plane of action is rotated through 90°. Hence for these in Fig. 7.16,

out-of-plane

$$I = \frac{bd^3}{12} = \frac{4 \times 0.021^3}{12} \times 7 = 2.2 \times 10^{-5}$$

$$C = \frac{bd^3}{6} = \frac{4 \times 0.021^3}{6} \times 7 = 4.4 \times 10^{-5}$$

$$A = bd = 4.0 \times 0.021 \times 7 = 0.59;$$

in-plane

$$I = \text{say}, \frac{(\text{span})^3 d}{12} = 600$$

$$A = bd = 4.0 \times 0.021 \times 7 = 0.59.$$

7.5.2 Interpretation of downstand grillage output

Figure 7.17 shows the element of beam of Fig. 4.3 but with interbeam shear flows r_{01} and r_{12} and balancing axial tension P. The equilibrium of the element differs in two ways from equation (4.1). Firstly, on resolving longitudinally,

$$\frac{dP_x}{dx} + (r_{12} - r_{01}) = 0$$

and secondly on taking moments about O_y,

$$\frac{dM_x}{dx} + \frac{z_c dP_x}{dx} = S_x \tag{7.7}$$

where z_c is distance of centroid of beam below centroid of slabs.

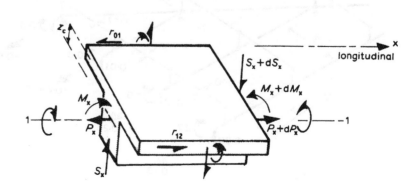

Fig. 7.17 Element of beam of beam-and-slab deck.

The axial stress at any level is calculated as for prestress beam from

$$\sigma = \frac{M_x z}{I_x} + \frac{P_x}{A_x}. \tag{7.8}$$

Figure 7.18 gives an example of the downstand grillage output for longitudinal and transverse members at a cross-section of a deck. The axial (bending) stresses derived from this output using equation (7.8) are shown in Fig. 7.19.

The web shear forces and beam torsions in longitudinal members are precisely as output.

The shear flow in the flanges (ignoring interbeam shear for the moment) is found as outlined for a beam in Section 2.3.2, except that the change in tension along the fibre element of Fig. 2.6 now includes a component due to gross axial force P in the beam;

$$r_f = \tau t = \frac{dM_x}{dx} \frac{A\bar{z}}{I_x} + \frac{dP_x}{dx} \frac{A}{A_x} \tag{7.9}$$

where $\bar{z}$ is the distance of centroid of element area A of cross-section

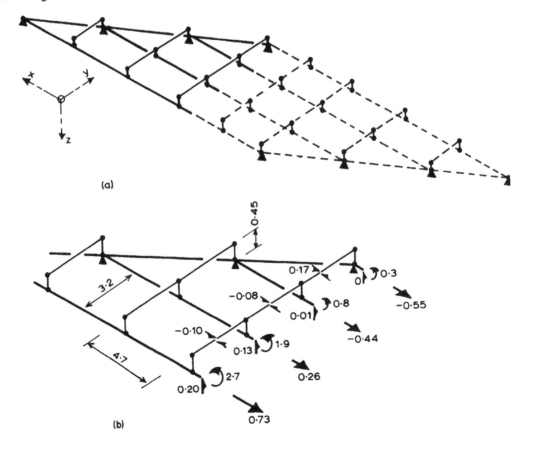

Fig. 7.18 Downstand grillage of deck with section of Fig. 7.19.

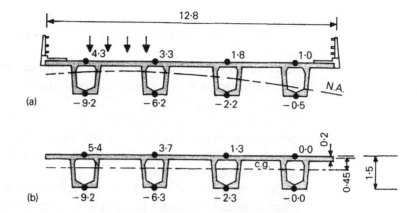

Fig. 7.19 Axial (bending) stresses in spaced beam-and-slab deck: (a) from downstand grillage and (b) from plane grillage.

from centroid of section. A_x and I_x are the area and second moment of area of the beam.

Using equation (7.7) we obtain

$$r_{\mathrm{f}} = \tau t = \frac{S_x A \bar{z}}{I_x} - (r_{12} - r_{01})\left[\frac{A}{A_x} - \frac{z_c \bar{z} A}{I_x}\right]. \quad (7.10)$$

The interbeam shear flows r_{12} and r_{01} can be calculated directly from the output by dividing the in-plane shear forces in transverse members in Fig. 7.18 by the width of member each represents. They are shown in Fig. 7.20(a) plotted as dashed lines. Using equation (7.10) and the output web shear forces S_x, we can calculate the flange shear flows at the edges of the web. The flange shear flows have triangular distributions decreasing from maximum values at the webs to zero at the edges of the flanges of the T-beams as shown in Fig. 7.20(b). They are shown in Fig. 7.20(a) superimposed on the interbeam shear flow to give the total shear flow picture.

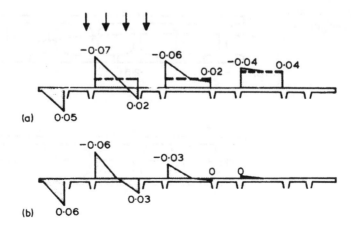

Fig. 7.20 Slab shear flows in spaced beam-and-slab deck: (a) from downstand grillage and (b) from plane grillage.

7.6 EFFECTS OF SLAB MEMBRANE ACTION ON BEAM-AND-SLAB DECK BEHAVIOUR

7.6.1 Axial stresses and movement of neutral axis

The axial stress at any point is a combination of stress due to bending moment M_x and tension P_x. Figs 7.19(a) and 7.21(a) show the axial stresses on the cross-sections of decks with beams spaced and contiguous, respectively. The neutral axis, where the combined bending and direct stress is zero, moves up in regions of deck subjected to load and downwards elsewhere.

For comparison, Figs 7.19(b) and 7.21(b) show the stresses calculated from bending moments in a plane grillage analysis. It is interesting to note that it is only the slab stresses which are significantly affected by slab membrane action. The soffit stresses are similar from the two analyses. However if the deck also has continuous upstand parapets, the movement of the neutral axis is much more significant at the edges, as discussed in Chapter 8.

7.6.2 Slab shear flows

Figure 7.20(b) shows the slab shear flows calculated from the plane grillage corresponding to the downstand grillage of (a). It is evident that for this spaced beam deck the interbeam shear flow is small and so the maximum in-plane shear from the plane grillage does not differ significantly from the downstand grillage. However this is not the case

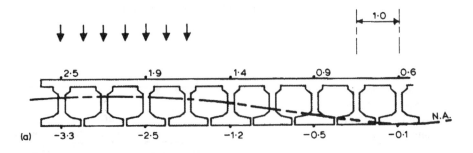
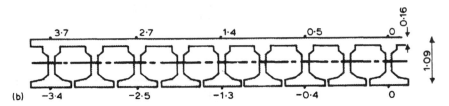

Fig. 7.21 Axial (bending) stresses in contiguous beam-and-slab deck: (a) from downstand grillage and (b) from plane grillage.

Effects of slab membrane action on beam-and-slab deck behaviour 175

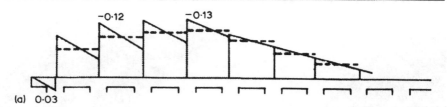

Fig. 7.22 Slab shear flows in contiguous beam-and-slab deck: (a) from downstand grillage and (b) from plane grillage.

for contiguous beam-and-slab decks, as shown in Fig. 7.22(a) and (b) for downstand and plane grillage, respectively. Because the beams are so close, the slab strips between are very stiff in shear, and as a result the interbeam shear flow forms a high proportion of the total. In the case of Fig. 7.22, the shear flow in the downstand grillage is three times that calculated from the plane grillage.

An approximate estimate of the high shear flow in the slab of contiguous beam-and-slab decks can be obtained from plane grillage output for the usually critical section at the edge of the loaded area. Slab membrane action, as discussed in Section 7.6.1, only has significant effect on the slab and not on the bottom flanges. As far as the beam at the edge of the loaded area is concerned, the load-free deck to the side behaves in membrane action like a very wide flange as shown in Fig. 7.23. The effective width of this flange is reduced by shear lag to about 1/10 to 1/6 span, as explained in Chapter 8. By recalculating the section properties of the deck under the load with such wide flanges and using the plane grillage output S_x (which differs little from S_x of downstand grillage), simple beam theory, equation (2.4), can be used to predict values of shear flow in the slab similar to those from the downstand grillage. It can be seen that the wider the beam spacing, the less significant the 1/6 span is compared to the flange associated with each beam, and thus slab membrane action has less significance.

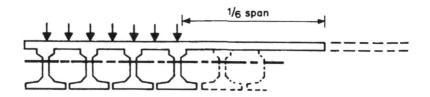

Fig. 7.23 Effective flange section for slab shear flow at edge of load.

Fig. 7.24 IH-10/Sam Houston Toll Road Interchange, Houston, Texas. Steel plate girders under long curved spans, pretensioned concrete beams under shorter spans. Designed by Brown & Root USA, Houston. Photograph E.C. Hambly.

REFERENCES

1. McHenry, D. (1943) 'A lattice analogy for the solution of stress problems', *J. Inst. Civ. Eng.*, **21**, pp. 59–82.
2. British Standard BS5400: Part 3: 1982, 'Steel, concrete and composite bridges, Part 3 Code of Practice for Design of Steel Bridges', British Standards Institution.

8 Shear lag and edge stiffening

8.1 SHEAR LAG

The thin slabs of cellular and beam-and-slab decks can be thought of as flanges of I- or T-beams as shown in Fig. 8.1. When such I- or T-beams are flexed, the compression/tension force in each flange near midspan is injected into the flange by longitudinal edge shear forces, shown in Fig. 8.2. (This figure also shows the coexistent transverse in-plane forces which prevent the flanges on each side of a web flexing away from each other.) Under the action of the axial compression and eccentric edge shear flows, the flange distorts (as in Fig. 8.3) and does not compress as assumed in simple beam theory with plane sections remaining plane. The amount of distortion depends on both the shape of the flange in plane and on the distribution of shear flow along its edge. As is evident in Fig. 8.3(a), a narrow flange distorts little and its behaviour approximates to that assumed in simple beam theory. In contrast, the wide flanges of (c) and (d) distort seriously because the compression induced by the edge shears does not flow very far from the loaded edge, and much of each wide flange is ineffective. The decrease in flange compression away from the loaded edge due to shear distortion is called 'shear lag'.

8.2 EFFECTIVE WIDTH OF FLANGES

To enable simple beam theory to be used for analysis of beams with wide flanges, the flanges are attributed 'effective flange widths.' The effective width of a flange is the width of a hypothetical flange that

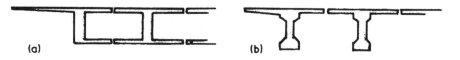

Fig. 8.1 Deck sections divided into I- or T-beams.

178 Shear lag and edge stiffening

Fig. 8.2 Forces on flange.

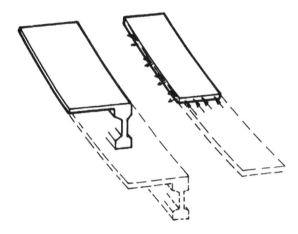

compresses uniformly across its width by the same amount as the loaded edge of the real flange under the same edge shear forces. Alternatively, the effective width can be thought of as the width of theoretical flange which carries a compression force with uniform stress of magnitude equal to the peak stress at the edge of the prototype wide flange when carrying the same total compression force. Figure 8.3 shows foam models of four shapes of flange under edge compression alongside diagrams of the equivalent effective flange under uniform compression. The figure shows, for each foam model, the non-linear end

Fig. 8.3 Shear lag distortion of flanges of various widths.

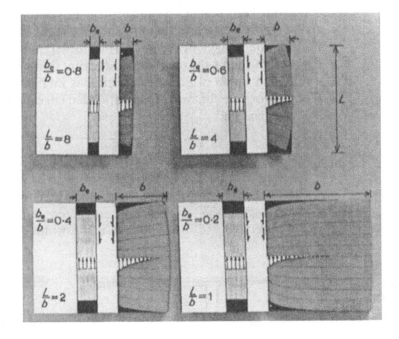

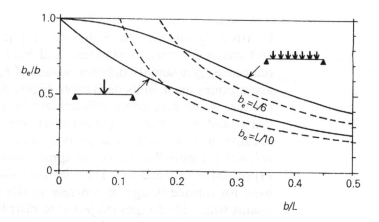

Fig. 8.4 Effective flange width ratio related to flange width-to-length ratio, for distributed and concentrated loading.

displacements and non-linear stresses at midlength, together with the effective flange width ratio b_e/b and length/width ratio L/b. It is evident that there is an upper limit to the effective flange width b_e however much the actual width b is increased.

Figure 8.4 illustrates approximate relationships between effective flange width ratio b_e/b and flange shape b/L, where L is the effective length between points of zero moment (contraflexure). The diagram shows different lines for a beam supporting a uniformly distributed load and a beam supporting a point load. The difference between these lines is explained in Fig. 8.5. Under the action of a point load in (a), the flange edge shear flow is large right up to the load, and the compression induced by shear flows near midspan cannot spread far across the flange. In contrast, the shear flows for the distributed load in (b) are predominantly applied near the ends, and the compression they induce has most of the length of the flange to spread out.

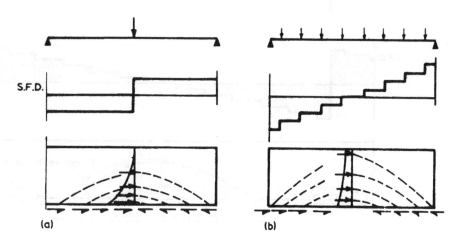

Fig. 8.5 Different stress distributions in flange for (a) concentrated load and (b) distributed load.

A rigorous analysis of the effects of load distribution and flange shape is extremely complicated. References [1]–[5] provide formulae, tables and charts for various structure and load conditions. The different references give slightly different values of b_e/b, as a result of different assumptions made in its derivation. Figure 8.4 should not be treated as precise. Some references differentiate between flanges between two webs, and edge flanges (cantilevers); this is not shown in Fig. 8.4 because the differences in b_e are only about 2% at $b/L = 0.1$ and 5% at $b/L = 0.3$. Figure 8.4 shows two approximate lines for effective widths expressed as $L/10$ and $L/6$. The effective width of $b_e = L/10$ is commonly used for routine design of concrete sections. This is conservative for beams with wide flanges subjected to distributed loads, where a value nearer $L/6$ may be valid. However $b_e = L/10$ is not conservative for beams with b/L less than 0.2 or near large concentrated loads, such as reactions at supports for continuous beams.

The calculation of effective flange widths of thin steel plates is more complicated than for concrete sections because the effectiveness of a thin flange is reduced by plate buckling as well as by shear lag. Naturally, if a detailed three-dimensional analysis is carried out using a lattice, folded plate or finite element model which considers slab membrane behaviour, the effects of shear lag are automatically included and it is not necessary to determine what the effective width of a flange is.

Fig. 8.6 Varying effective width of flange of continuous deck.

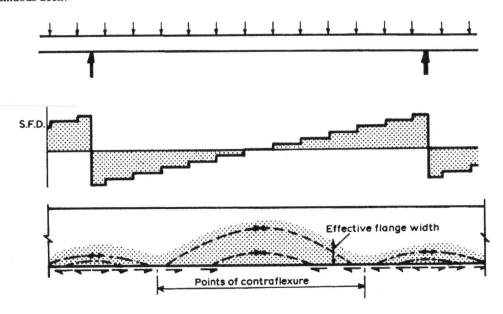

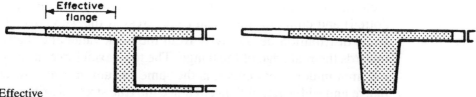

Fig. 8.7 Effective sections for grillage analysis.

Continuous bridge decks have significant variations in effective flange width along their length. This is partly due to the different distances between points of contraflexure along spans and over supports, and partly due to the very different dominating loads distributed over spans and concentrated at support reactions. Figure 8.6 demonstrates the variations of edge shear flow along the length of a deck, and of spread of compression or tension across the flange.

The effects of shear lag are incorporated in grillage analysis by reducing the flanges associated with each longitudinal beam in accordance with Fig. 8.4. Figure 8.7 shows part cross-sections of two decks and the effective sections used for calculation of grillage member properties. These sections are also used for calculating from the grillage output the values of the peak bending stresses at webs and the values of the shear flows at the roots of the flanges.

The true distribution of longitudinal bending stress in a wide flange decreases towards the outside edge. Its shape can be estimated from the peak value, derived from grillage, and the effective width of the flange. Figure 8.8 shows the effective section for grillage of an elemental I-beam of a cellular deck. Applying simple beam theory to the reduced

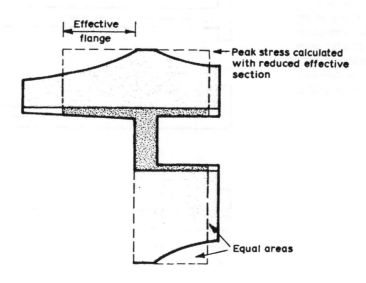

Fig. 8.8 Determination of bending stress distribution from peak value and effective flange width.

section, the stress is uniform across each effective flange (as shown dotted) and equal to the actual peak stress at the web edge. The true stress distribution decays away from the peak value and 'flattens out' towards the real edge of the flange. The total axial force in each flange, i.e. area under stress curve, is the same for uniform narrow effective flange and wider actual flange. Hence the true stress distribution can be sketched to pass through the peak value and enclose the same area as the uniform stress on the reduced flange.

Some designers ignore shear lag in the global grillage analysis and only consider it in the detailed design of the section. In calculations for working stress conditions, this approach has the disadvantage that the sections most affected by shear lag are made over-stiff and so attract high moments which have to be carried by the section in the detailed design. If shear lag is considered consistently in the global and local analyses the distribution of stiffnesses between span and support regions is more realistic. Then the sections most affected by shear lag shed load to other regions and are not penalized in the detailed design. This is demonstrated in Section 4.8.2. However in calculations for collapse conditions it may be appropriate to ignore shear lag, in the global and local analyses, if the flanges yield so that the most highly stressed regions near the webs shed load to the outer edges.

8.3 EDGE STIFFENING OF SLAB DECKS

A slab deck is better able to carry a load near an edge if the edge is stiffened with a beam. Figure 8.9(a) shows a slab deck with edge

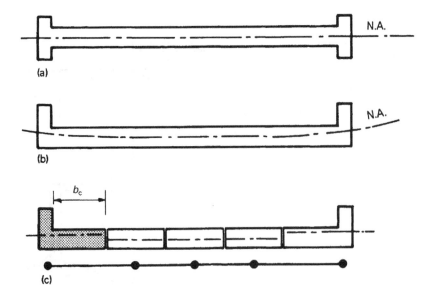

Fig. 8.9 Edge stiffening of slab: (a) edge beam centroids on midplane of slab; (b) edge beam centroids above midplane of slab; and (c) sections for grillage.

stiffening beams which have their centroids on the midplane of the slab. The bending inertias of such beams are calculated about the midplane of the slab and the beam sections are fully effective. Improved edge stiffening is achieved if the beams do not have their centroids on the midplane of the slab as in (b) because the beams then act as L-beams with the slab deck acting to some extent as a flange. Under bending action, the neutral axis remains near the midplane of the slab in central regions and rises towards the edges. The width of the slab that acts as flange to the edge beam is restricted by the action of shear lag. The effective width can be determined as described in Section 8.2.

8.4 UPSTAND PARAPETS TO BEAM-AND-SLAB DECKS

The load distribution characteristics of a beam-and-slab deck can be greatly improved by making the parapet part of the structure. Figure 8.10(a) and (b) show the bending stresses computed from a folded plate analysis (see Chapter 12) of decks with and without upstand parapets supporting loads near the edge beams. It is evident that while the top of the upstand attracts a high compressive stress, the accompanying stresses in the edge main beam are much smaller than those in the deck without structural parapet. The parapet effectively acts with the edge main beam, as shown in Fig. 8.11.

The predicted slab stresses in Figs 8.10 and 8.11 differ because in the grillage analysis the slab to the right of the loaded beam is not subjected to additional compression by interbeam shear as described in Chapter 7.

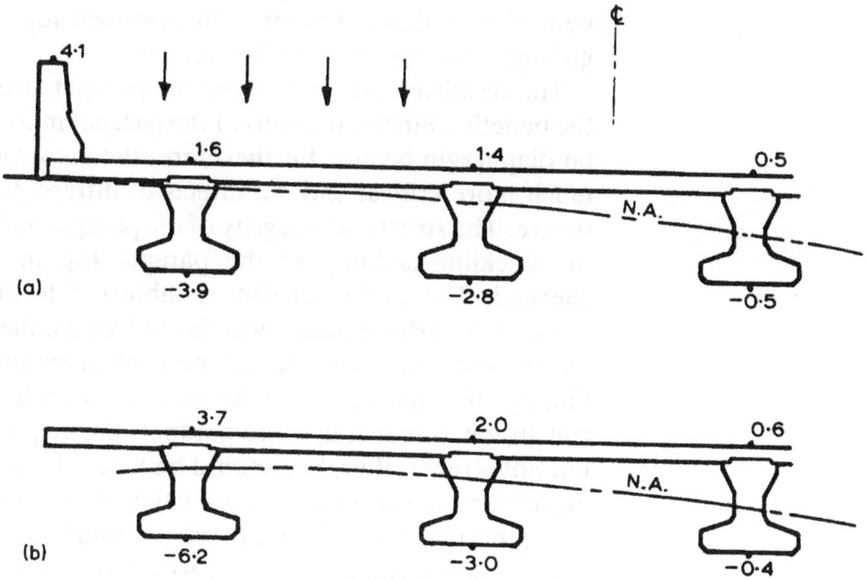

Fig. 8.10 Bending stresses in beam-and-slab deck: (a) with and (b) without structural parapet. (From folded plate analysis.)

Fig. 8.11 Bending stresses predicted by grillage for deck for Fig. 8.10.

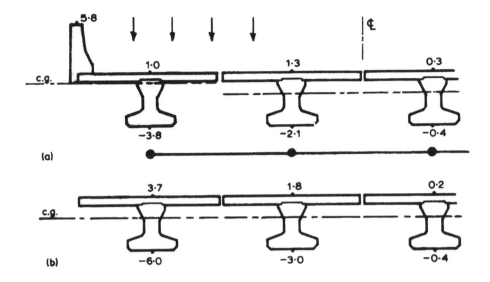

Furthermore, but in contrast, by assuming in the grillage that the parapet is part of the edge beam, shear lag deformation of the thin cantilever slab has been ignored and the parapet appears more effective, thus attracting higher stresses. Such comparison between a plane grillage and a three-dimensional structural analysis is highly dependent on the closeness of the beams. If the beams are close together, several of them will act compositely with the upstand parapet. Consequently, if the parapet is assumed part of the structural edge beam in a plane grillage analysis, it is advisable to check the effectiveness of the section by comparing the level of its centroid with the level of the neutral axis derived from a three-dimensional analysis (downstand grillage, folded plate or finite element).

The disadvantages of making the parapet structural often outweigh the benefits. Firstly, the ends of the parapet must be properly supported on diaphragm beams. Furthermore, the construction sequence can be much more critical and the effects of differential shrinkage are more severe. The structural integrity of the parapet and joints is essential, and the buckling stability of the parapet top in compression must be checked. The cantilever slab is subjected to longitudinal shear flow forces two or three times those in the free cantilever. The parapet must be so strong that it could not be broken by the impact of a vehicle. Finally, the analysis must be carried out with much more care. A conclusion is that if upstand parapets are required for traffic reasons, but not structurally, they should be made discontinuous with frequent expansion joints (at the same time they must retain sufficient transverse strength to prevent vehicles passing through).

Upstand parapets can also be used structurally on cellular decks, but the disadvantages are generally even more pronounced. The load distribution characteristics of a cellular deck are usually so good that the whole cross-section, or a wide part of it, is effective in supporting a load near an edge. The addition of a tall upstand has little effect on stiffness while it attracts high compressive stresses to the top of the upstand and high shear stresses to the cantilever slab. However, a small beam on the outside edge of a wide cantilever can be useful in providing local stiffening to the edge.

8.5 SERVICE BAYS IN BEAM-AND-SLAB DECKS

Services are often carried on bridges in service bays placed under the footways or verges. Some are open at the top along their full length (but spanned by paving slabs), as shown in Fig. 8.12(b), while others are effectively box beams as in Fig. 8.12(c). The structural actions of these different structures when loads are placed near them are very different. Figure 8.12 shows the bending stresses computed from folded plate analysis of the three different structural forms. It is evident that the best distribution (i.e. with lowest maximum stresses) is obtained when the edge beams behave as a box. However, the shear force in the web next to the edge is then large and often critical for design, since in addition to attracting a high bending shear it is subjected to high torsional shear

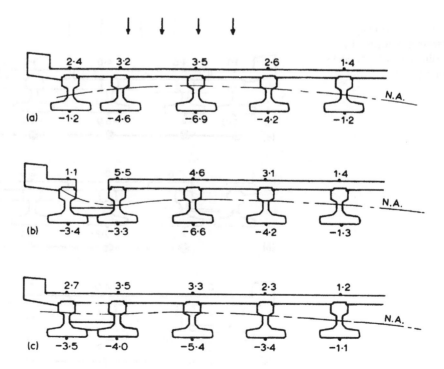

Fig. 8.12 Bending stresses in beam-and-slab deck with various edge details: (a) without service bay; (b) open service bay; and (c) box service bay. (From folded plate analysis.)

flows. It should be noted that the box will only be torsionally stiff if the bottom slab is effectively connected to the web on each side for transfer of longitudinal shear flow. In practice, it is very difficult to construct a structurally effective *in situ* concrete slab to form the bottom of the trough or box between precast beams, as the joints are unlikely to have sufficient stiffness in either bending or longitudinal shear.

The deck with an open trough service bay has the worst distribution because the trough behaves as a U-beam subjected to eccentric load and the outside web is largely ineffective. This inefficiency can be largely avoided while retaining the benefit of easy access by giving the service bay the box cross-section near midspan and the open trough section towards the ends. The high transverse bending stiffness of the box at midspan provides effective load distribution where it is needed, while the removal of the top slab near the supports avoids high torsions in the edge box which overstress the web next to the edge in shear. Finally, it should be noted that if the stiffness and strength of the outside beam is ignored in the design, the slabs connecting this beam to the deck should be designed to articulate. If they are made continuous but only given nominal strength, they will break when the deck flexes.

Figure 8.13 shows, for comparison, the bending stresses computed from various grillage models of the decks in Fig. 8.12. The different sections assumed for the edge beam are shown.

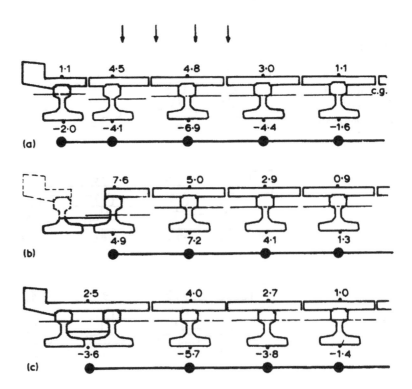

Fig. 8.13 Bending stresses predicted by grillage models of decks of Fig. 8.12.

Fig. 8.14 Eleven-span flyover at interchange of Highways 401 and 410, Toronto, Canada; concrete structure cast-in-place and post-tensioned in five longitudinal stages. Designed by the Structural Office of the Ontario Ministry of Transportation. Photograph courtesy of Ontario Ministry of Transportation.

REFERENCES

1. American Association of State Highway and Transportation Officials (AASHTO) 'Standard Specifications for Highway Bridges', Washington DC, 14th edn, 1989.
2. British Standard BS5400: Part 3: 1982, 'Steel, concrete and composite bridges, Part 3 Code of Practice for Design of Steel Bridges', British Standards Institution.
3. Ontario Highway Bridge Design Code, with Commentary 1990, Ontario Ministry of Transportations and Communications, 3rd edn.
4. Roark, R.J. and Young, W.C. (1989) *Formulas for Stress and Strain*, McGraw-Hill, New York, 6th edn.
5. Nakai, H. and Yoo, C.H. (1988) *Analysis and Design of Curved Steel Bridges*, McGraw-Hill, New York.

9   Skew, tapered and curved decks

9.1 SKEW DECKS

9.1.1 Characteristics of skew decks

The majority of bridge decks built today have some form of skew, taper or curve. Because of the increasing restriction on available space for traffic schemes and also due to the increasing speed of the traffic, the alignment of a transport system can seldom be adjusted for the purpose of reducing the skew or complexity of the bridges. Fortunately, this increasing demand for high skew bridges has been accompanied by the development of computer-aided methods of analysis, and it is now generally possible to design a structure at any angle of skew.

In addition to introducing problems in the design of details of a deck, skew has a considerable effect on the deck's behaviour and critical design stresses. The special characteristics of skew of a slab deck are summarized in Fig. 9.1. They are:

1. variation in direction of maximum bending moment across width, from near parallel to span at edge, to near orthogonal to abutment in central regions;
2. hogging moments near obtuse corner;
3. considerable torsion of deck;
4. high reactions and shear forces near obtuse corner;
5. low reactions and possibly uplift in acute corner.

The size of these effects depends on the angle of skew, the ratio of width to span, and particularly on the type of construction of the deck and the supports. Figure 9.2 shows how the shape and edge details can influence the direction of maximum moments. While in (a) and (b) the decks span on to the abutments, in (c) the stiff edge beam acts as a line support for the slab which effectively spans right to the abutment across the full width. In (d) the skew is so high that the deck is cantilevered off the abutments at the acute corners.

Fig. 9.1 Characteristics of skew slab deck.

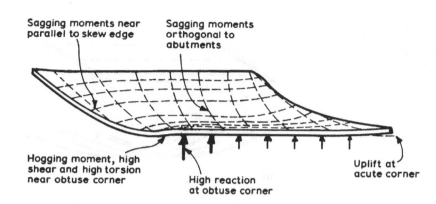

The deleterious effects of skew can be reduced by supporting the deck on soft bearings. The high reaction on the bearing at the obtuse corner is shed to neighbouring bearings. In addition to reducing the magnitude of the maximum reaction, this also reduces the shear stresses due to shear and torsion in the slab and it reduces the hogging moment at the obtuse corner. Uplift at the acute corner can also be eliminated. However, this redistribution of forces along the abutment is accompanied by an increase in sagging moment in the span.

The above characteristics are particularly significant in solid and cellular slab decks because their high torsional stiffness tries to resist the twisting of the deck. In contrast, skew is less significant in beam-and-slab decks, particularly with spaced beams. Figure 9.3(a), (b) and (c) shows a plan, elevation and right section of a spaced beam-and-slab

Fig. 9.2 Principal moment directions in skew slab decks.

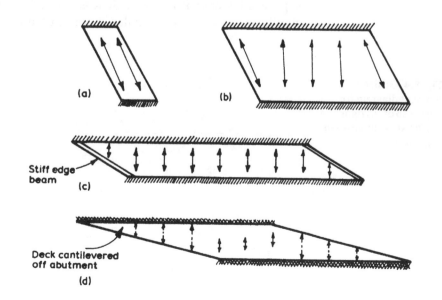

---

construction outweighs the increase in reinforcement involved. If a deck has a high skew angle with reinforcement parallel to the skew directions, the quantity of reinforcement is likely to be excessive and uneconomic. In general, the grillage members should be placed parallel to the designed lines of strength, which are usually orthogonal. Clark [3] shows that for narrow slab decks the reinforcement should be placed in directions of mesh of (b), while for decks with abutment length greater than span, the directions of (c) are generally more appropriate. However, for cellular decks and beam-and-slab decks, the longitudinal grillage members should be parallel to the webs or beams which are usually as in (b).

While the general comments of Chapters 3, 4 and 5 apply to skew decks, a problem arises in relation to member properties near the diaphragm beams in Fig. 9.5(b) and edge beams in (c). It might appear that if members are given uniform properties, the parts of the deck in the triangular regions are represented more than once. If the orthogonal members are assumed to represent the typical deck construction, they should have the typical section properties. No reduction in the grillage member stiffness is necessary since the strip of prototype it represents is not tapered and does not terminate at a point. The edge trimming grillage member should be given the stiffness appropriate to the additional stiffness coincident with its line in the prototype. If the prototype is a slab with no special end or edge stiffening, the skew grillage member is given nominal stiffness. However, if there is a stiffening diaphragm or edge beam built into or on to the deck, the equivalent grillage member should be given the stiffness appropriate to the dimensions and strength of the actual stiffening member with the part of slab that participates as its flange.

Even though the lines of design strength and grillage member may be chosen to be near parallel to the principal moment directions, the grillage might still predict high torques in places. The maximum moments and stresses in slab decks are then calculated from the orthogonal moments and torques by using the equations of Section 3.3.3. The reinforcement of concrete decks must be designed to resist the combinations of moment and torque, and the equations of Wood [4] and Armer [5] are relevant. The grillage output of beam-and-slab and cellular decks is interpreted in the ways described in Sections 4.7 and 5.7, respectively.

9.2 TAPERED DECKS

Bridge decks seldom have a very pronounced taper, and a tapered grillage as in Fig. 9.6 can be used without special thought. The only problem is that grillage member properties must be incrementally

190 Skew, tapered and curved decks

Fig. 9.3 Skew beam-and-slab deck: (a) plan; (b) elevation; and (c) section.

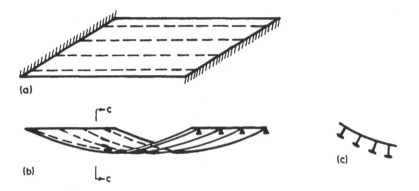

deck subjected to uniform load. At the abutments there is a large difference in longitudinal slope at adjacent points on neighbouring beams (evident in (b)) and also a relative vertical displacement (evident in (c)). This distortion of the deck can occur without generating large reactive forces if the torsional stiffnesses of the slab and beams are low. Under the action of a local concentrated load, distribution still takes place by transverse bending of the slab, but the beams behave much as in a right deck spanning longitudinally. However, the increase in beam shear force and reaction at the obtuse corner is still significant and should be considered. Uplift at the acute corner is unlikely. It should be noted that if the beams have box sections with high torsional stiffness, they will attract high torques. It may well be found that the torsion shear in the webs is then excessive, and torsionally flexible I-beams may be more appropriate.

The effects of skew are generally considered negligible for simply supported decks with skew angle less than 20°. However, the effects are significant at lower skew angles in continuous decks, particularly in the region of intermediate supports. Figure 9.4(a) and (b) shows grillage bending moment diagrams for the edge web of a three-span cellular deck with right supports in (a) and 20° skew in (b). Both are loaded on the centre span. There is little difference in midspan moment. However, at the support skewed towards the loaded span the moment, shear force (slope of saw tooth moment diagram) and reaction are greatly increased by the skew.

9.1.2 Grillage meshes for skew decks

Design moments in simply supported skew isotropic slab decks can be obtained from the influence surfaces of Rusch and Hergenroder [1] or Balas and Hanuska [2]. However, these charts have the disadvantage that they are difficult to use, do not give the user a complete picture of the force system in the deck under a particular load case, and cannot be used for orthotropic, cellular or beam-and-slab decks because of their very different distortional and torsional characteristics. In general, grillage analysis is much more convenient for all types of deck. Even during the preliminary design stage when it is not clear what span-to-depth ratio is appropriate to the skew and method of construction, a preliminary quick crude grillage is preferable to interpretation and conversion of the charts.

A skew deck can be analysed with a grillage having either a skew mesh as in Fig. 9.5(a) or orthogonal mesh as in (b) or (c). While the skew mesh is convenient for low skew angles, it is not appropriate for angles of skew greater than 20° because it has no members close to the direction of dominating structural action. However some designers use reinforcement and grillage members at skews up to 30° on beam-and-slab decks because they find the simplification of detailing and

Fig. 9.4 Grillage moment diagram for edge web of (a) right and (b) 20° skew three-span cellular decks.

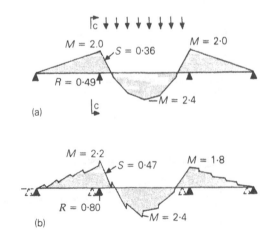

Fig. 9.5 Grillages for skew decks: (a) skew mesh; (b) mesh orthogonal to span; and (c) mesh orthogonal to support.

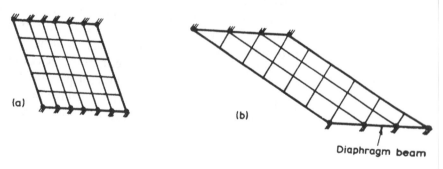
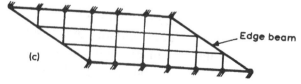

Fig. 9.6 Tapered grillage.

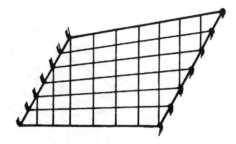

increased along the strings of members, thus making data preparation and output interpretation cumbersome. Often a small taper can be ignored in the analysis as it has little effect on the deck's behaviour. However, if a deck also is at a high skew angle, the edges will have very different spans and both taper and skew must be reproduced.

A fan-type structure, as Fig. 9.7, can be analysed with a grillage. The subtended angle between neighbouring radial members can be as large as 15° without introducing significant error. The grillage mesh should be near 'square', and the radial members given the stiffness equivalent to the section midway along their length. It might appear that the mesh should get very fine close to the centre but when thought is given to how the lines of strength in the prototype, in the form of reinforcement of beams, cannot taper to a point but must be curtailed, it will be evident that the grillage mesh should also be curtailed.

9.3 CURVED DECKS

9.3.1 Curved beams

When a vertical load is placed on a curved beam there is an interaction of moment and torsion along the length of the beam as explained in references [6] and [7]. This interaction of forces can be investigated much

Fig. 9.7 Grillage for fan slab.

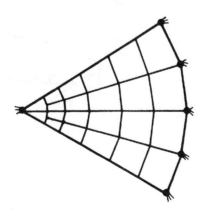

Fig. 9.8 Forces on element of curved beam.

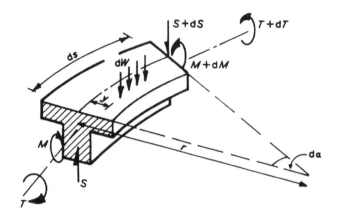

more simply with a computer analysis than by hand. A curved bridge can be analysed without difficulty with a space frame model, or grillage, in which curved members are represented by 'curved' strings of straight members as described below. In multibeam decks, interaction between moment and torsion is significant at low angles of curvature due to redistribution between beams. Before discussing grillage analysis of curved decks, the equilibrium and stiffness equations for a curved beam are first presented.

Figure 9.8 shows an element of beam curved in plan. It has length ds, radius of plan curvature r, and subtends an angle dα. It is subjected to an element of vertical load dW at eccentricity y; this force is resisted by the

Fig. 9.9 (a) End forces and (b) end displacements on curved beam.

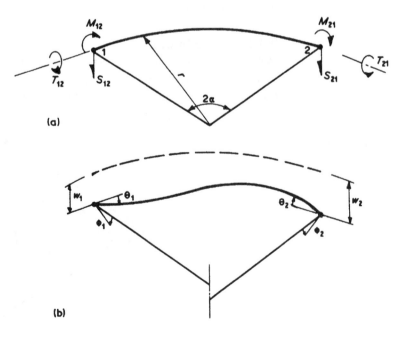

beam moment M, shear force S and torsion T. Equilibrium of the element requires:

$$dS = -dW = -W(s)\,ds$$
$$\frac{dM}{ds} - \frac{T}{r} = S \qquad (9.1)$$
$$\frac{M}{r} + \frac{dT}{ds} = y\frac{dW}{ds}.$$

The flexural and torsional stiffness of the element are the same as those for the straight beam in equations (2.5) and (2.14) except that moment and torsion are related to deflections by

$$M = -EI\left(\frac{d^2 w}{ds^2} - \frac{\phi}{r}\right) \qquad T = -GC\left(\frac{d\phi}{ds} + \frac{1}{r}\frac{dw}{ds}\right). \qquad (9.2)$$

If the flexural and torsional rigidities EI and GC are uniform along the beam, the end forces and end displacements of Fig. 9.9 are related by the complex slope–deflection equations given in Nakai and Yoo [6].

9.3.2 Moments and torsions along a curved deck

Figure 9.10 illustrates an isometric view of a space frame model for a continuous curved bridge deck with four spans of lengths 25 m, 31 m, 31 m and 25 m. The deck is supported by one bearing at each pier, with stability against overturning provided by the arrangement of the supports. Figure 9.11(a) and (b) illustrates the distribution of bending moments and torques along the spans due to: (i) a uniformly distributed load of 0.1 MN/m (shown continuous), and (ii) a patch load of 1 MN

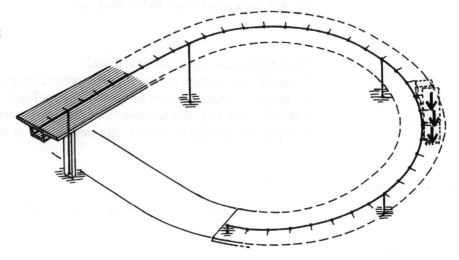

Fig. 9.10 Continuous beam space frame model for a four span curved bridge.

Fig. 9.11 (a) Moments and (b) torque along spans of curved bridge of Fig. 9.10 under distributed loading and eccentric concentrated loading.

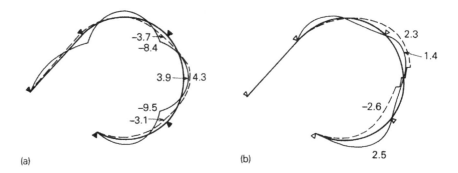

placed 2 m outside the centre line on span 2 (shown dashed). It is evident in Fig. 9.11(b) that at the ends of the deck there are no torques because there is only one bearing. It is also evident that due to the distributed load the torque is zero (i.e. reverses direction) near the centre of each span and near each support. However the torque due to the patch load passes the supports and decays until it reaches the end of the curve. It should be noted that since equations (9.2) involve deflection and twist in the formulae for both moment and torque, the distributions of moments and torques both depend on both of the section properties I and C.

The distributions of bending moments and torques in Fig. 9.11(a) and (b) can be compared with the moments and torques along a straight deck with similar cross-section and spans, which are shown in Fig. 9.12(a) and (b). In order to give the straight deck stability against overturning it is provided with two bearings at every other support (shown as solid triangles). As a result the torque due to the eccentric patch load is transmitted only to the nearest supports with torsional stiffness, as explained in Section 2.4. It is evident from Fig. 9.11(a) and Fig. 9.12(a) that the curved deck has to carry larger moments than the straight deck in addition to the torques.

9.3.3 Grillage analysis of curved decks

A curved bridge deck can be represented for the purpose of analysis by a grillage composed of curved members as in Fig. 9.13(a), or of straight members as in (b). While some computer programs do have the facility to represent curved members, the improvement in accuracy over the

Fig. 9.12 (a) Moments and (b) torques along straight bridge for comparison with curved bridge in Fig. 9.11.

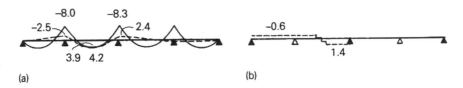

Fig. 9.13 Grillages of curved decks: (a) curved members and (b) straight members.

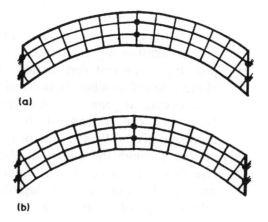

Fig. 9.14 Curved concrete box-girder Kylesku Bridge, Sutherland, Scotland; designed by Ove Arup & Partners. Photograph courtesy Ove Arup & Partners.

straight member grillage is not significant enough to warrant its general use. Most programs do not have the facility and straight members must be used in any case.

It will be found that if the general recommendations of Chapters 3, 4 and 5 are used to determine member spacing, the maximum change in

direction at a joint will seldom need to be more than 5°. This is much smaller than the angle at which the behaviour of a true curved beam differs significantly from the 'curve' of straight members. In the prototype, moment and torsion interact continuously and smoothly along a curved member. In the straight member grillage such interaction only occurs at joints, so that each type of force is discontinuous. However, the values of forces midway along members are representative of those in the prototype. A smooth distribution can be plotted through the values at these points, and from it the values elsewhere interpolated. The interpretation of the grillage output is precisely the same as that described in Chapters 3, 4 and 5 for the relevant type of construction.

REFERENCES
1. Rusch, H. and Hergenroder, A. (1961) *Influence Surfaces for Moments in Skew Slabs*, Munich, Technological University. Translated from German by C.R. Amerongen. London, Cement and Concrete Association.
2. Balas, J. and Hanuska, A. (1964) *Influence Surfaces of Skew Plates*, Vydaratelstvo Slovenskej Akademie Vied, Bratislava.
3. Clark, L.A. (1970) 'The provision of reinforcement in simply supported skew bridge slabs in accordance with elastic moment fields', Cement and Concrete Association, Technical Report.
4. Wood, R.H. (1968) The reinforcement of slabs in accordance with a predetermined field of moments, *Concrete*, **2**, 69–76.
5. Armer, G.S.T. (1968) Discussion of Wood (1968), *Concrete*, **2**, 319–20.
6. Nakai, H. and Yoo, C.H. (1988) *Analysis and Design of Curved Steel Bridges*, McGraw-Hill, New York.
7. American Association of State Highway and Transportation Officials (AASHTO) (1980) 'Guide specifications for horizontally curved highway bridges', Washington DC.

10 Distribution coefficients

10.1 INTRODUCTION

Prior to the general use of computer methods, charts of distribution coefficients provided the most convenient method of load distribution which was quick enough for general design use. Finite difference relaxation methods by hand computation were available but were too cumbersome for routine designs. Numerous charts have been published for hand calculation of critical design moments, etc. in isotropic and orthotropic slabs and other types of deck. Computer programs are also available, such as in Jaeger and Bakht [1], which simplify this approach to load distribution analysis. This chapter reviews some of the previously published charts and then demonstrates the use of three charts for preliminary design of slab, beam-and-slab and cellular decks.

All charts have limitations. In general, a chart which enables direct calculation of design moment is restricted to an individual type of deck construction subjected to a particular load case, which may not be the most critical. In contrast, a more versatile set of charts which enable calculation for a variety of deck types under a variety of load cases is more difficult to use and requires a considerable amount of interpolation and hand calculation to determine the critical conditions. Often the complexity of instructions discourages the user from taking the trouble.

Many of the charts available enable accurate calculations to be made for longitudinal bending moment in simply supported bridge decks. However, in general, they are not able to represent in detail the wide variety of cross-section construction, and predictions of transverse moments are often unsatisfactory for all but simple slab decks. In addition, few charts enable accurate calculation of shear force at supports, which usually have a different distribution from that of longitudinal bending moments.

The three charts demonstrated in Sections 10.3 to 10.6 have been designed to permit rapid investigation of the load distribution characteristics of a wide variety of types of deck construction, and to enable rapid calculation of the maximum design moments. To make the charts versatile and simple, no explanation is given of differences in secondary characteristics of the various types of deck. Such behaviour can be highly dependent on factors such as skew, continuity and cross-section type, which can all be represented much more satisfactorily in a grillage or other computer-aided analysis. These charts are solely intended to help the designer make the initial choice of type of construction and deck dimensions.

10.2 SOME PUBLISHED LOAD DISTRIBUTION CHARTS

10.2.1 Isotropic slabs

One of the most useful sets of design charts is the book of influence surfaces for isotropic slabs with various shapes and support conditions produced by Pucher [2]. In addition to being useful for the determination of critical design moments in simply supported right slab decks, these charts also provide one of the simplest methods of determining moments under concentrated loads on secondary slabs of beam-and-slab and cellular decks. For both applications, the presentation of the influence surfaces makes it relatively easy to isolate the critical load position and then calculate design moments.

Skew simply supported isotropic slabs can be analysed by means of the charts of Rusch and Hergenroder [3] or of Balas and Hanuska [4]. While a great deal of valuable information can be derived from these charts for particular shapes of deck, interpolation between charts of different aspect ratios and different skews can be extremely cumbersome and confusing. In general, even for preliminary design, it is quicker and more reliable to carry out a quick grillage analysis, as described in Chapter 9, with estimated properties.

10.2.2 Orthotropic slabs

The most widely used charts for load distribution in orthotropic slab decks are those of Morice and Little [5]. The theoretical basis of these charts is described by Rowe [6] who describes and demonstrates their application in detail. There are two series of charts which give the load distribution of slabs having no torsional stiffness and for slabs having the full torsional stiffness of isotropic slabs. For most bridge decks, interpolation between the sets of charts is necessary and the references demonstrate a simple though somewhat lengthy tabulated procedure.

The above charts, which are based on harmonic analysis demonstrated in Chapter 12, give the distribution of deflections due to the first harmonic of load. Fortunately, the distribution of longitudinal moments due to most design loads approximates closely to that of the first harmonic deflections. However, transverse moments are highly dependent on the local distribution (i.e. higher harmonics) of concentrated loads, and it is necessary to superpose several harmonic components of the transverse moments. The references describe a second more complicated tabular procedure for this analysis.

To avoid the necessity of superposing a number of harmonic components while determining transverse moments, Cusens and Pama [7] developed a set of charts similar to those of references [5] and [6], but with the first nine harmonics of a midspan point load already superposed. While this reduces the quantity of computation for transverse moments it means that the charts specifically apply to loads near midspan. This probably does not matter for the purpose of preliminary design and these charts have been found very convenient by some designers. Furthermore, they have the additional advantage of covering a wider range of torsional stiffnesses which enables them to be used for shear-key decks as described in Chapter 6. Shear-key decks can also be analysed with the charts of reference [8] which were derived specifically for them.

The charts of references [4]–[8] all relate to orthotropic slab decks. They can also be used accurately for beam-and-slab and cellular decks if the decks' concentrated stiffnesses can be notionally 'spread out' into a continuum without changing the decks' characteristics. In other words, the charts assume that the deck cross-section deflects in a smooth curve as shown in Fig. 10.1(a). If the deck cross-section deflects as in Fig. 10.1(c) with 'steps' at the concentrations of stiffness, the continuous charts cannot simulate such distortion of the cross-section. This limitation does not apply to the charts demonstrated in the following

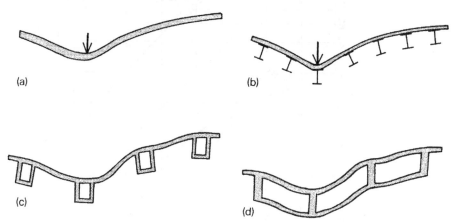

Fig. 10.1 Cross-sections of various decks: (a) and (b) distorting in smooth curve; (c) and (d) distorting in series of steps.

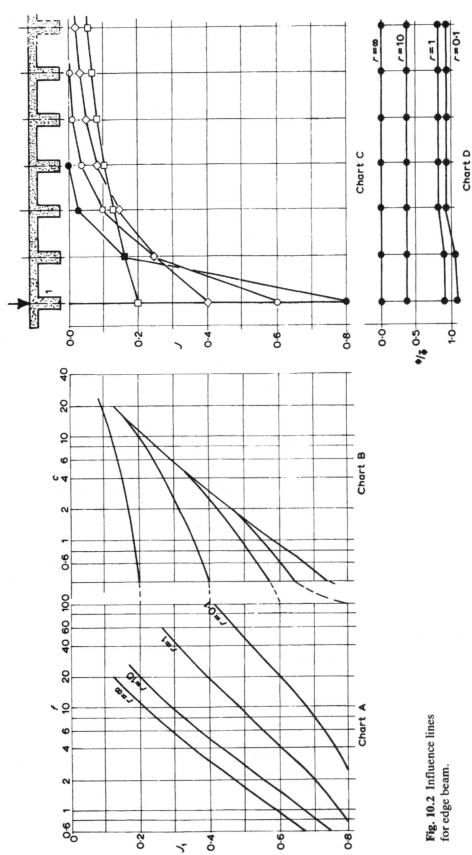

Fig. 10.2 Influence lines for edge beam.

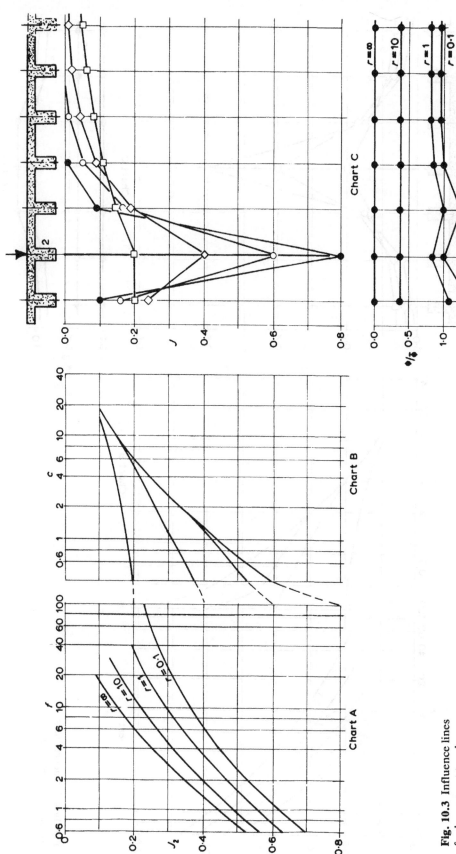

Fig. 10.3 Influence lines for beam next to edge.

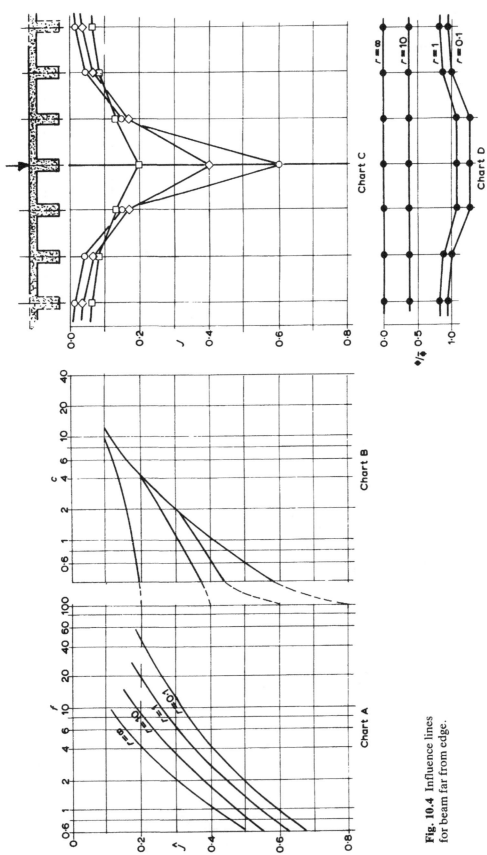

Fig. 10.4 Influence lines for beam far from edge.

sections since they are based on the assumption that stiffnesses are concentrated at points across the cross-section so that the deck can distort in 'steps' if appropriate.

10.3 INFLUENCE LINES FOR SLAB, BEAM-AND-SLAB AND CELLULAR DECKS

Figures 10.2–10.4 contain charts for the determination of influence lines for various points across the cross-section of simply supported right decks of slab, beam-and-slab and cellular construction. The charts were developed from repeated application of the approximate folded plate method outlined in Appendix B of reference [9].

The first step in the analysis of any deck with these charts is to notionally subdivide the deck into a number of parallel 'beams' as shown in Fig. 10.5 and as is necessary for grillage analysis. The physical characteristics of the deck can then be summarized by three non-dimensional parameters which relate the various stiffnesses of the structure. The parameters are as follows.

f, the flexural stiffness ratio. This relates the transverse flexural stiffness of the slab or slabs between 'beams' to the longitudinal flexural stiffness of the 'beams'.

r, the rotational stiffness ratio. For slab and beam-and-slab decks this relates the torsional stiffness of the slab and beams to the transverse flexural stiffness of the slab. For cellular decks, r relates the flexural stiffness of the webs to the transverse flexural stiffness of the slabs.

c, the cellular stiffness ratio. This applies only to cellular decks and relates the cellular torsion stiffness of the deck to the longitudinal flexural stiffness of the 'beams'.

Because the basic equations for slab, beam-and-slab and cellular decks are similar, it has been possible to derive a single set of charts for all three types of deck. Since the dominating terms in f and r are different for different types of construction, their algebraic definitions

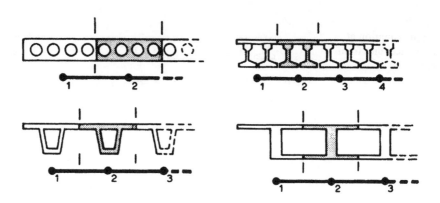

Fig. 10.5 Deck cross-section divided into 'beams'.

are given separately in the demonstrations of the charts in Sections 10.4–10.6.

The assumptions of the approximate folded plate theory are given in reference [9]. Some of these are listed below, together with additional assumptions necessary to simplify application of the charts.

1. The deck is right and simply supported. The charts can also be used for continuous decks for the lengths between points of contraflexure on each span.
2. The deck is prismatic; i.e. it has the same section from end to end. (If the deck section does vary, the properties near midspan are the most appropriate for approximate use of the charts.)
3. Slabs and beams are prevented from rotating about a longitudinal axis at their ends by rigid diaphragms.
4. There are no midspan diaphragms, or the stiffness of such diaphragms is assumed to be 'spread out' along the deck.
5. The deck has the same transverse bending stiffness at every point across the section. Hence for the beam-and-slab decks of Fig. 10.1 it is assumed that the slab spans transversely between centre lines of 'beams', at which the bending and torsional stiffnesses are assumed concentrated. In the same way, for cellular decks it is assumed that both top and bottom slabs are continuous across the deck and only connected to webs at their midplanes. (If the charts are to be used to analyse a spaced beam-and-slab deck constructed of wide box beams which distort, a supplementary plane frame analysis of the cell distortion is necessary to determine the equivalent uniform slab.)
6. The deck cross-section can be divided into a number of identical equidistant 'beams'. Edge stiffening can only be represented if it can be thought of as an additional width of the uniform or repetitive section.
7. The centroid of the deck is everywhere at the same level, so that, as in plane grillage analysis, slab membrane action of beam-and-slab decks is ignored.

Fig. 10.6 Approximation of influence lines from charts.

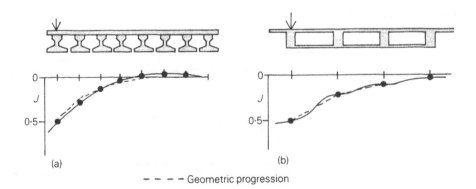

– – – – Geometric progression

Fig. 10.7 Definition of 'beam' rotation ϕ and average rotation $\bar{\phi}$.

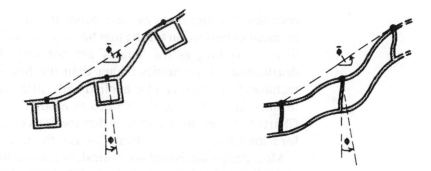

8. In cellular decks the ratio of the individual flexural stiffnesses of the top and bottom slabs is within the range 0.3–3. (If the slabs differ by more than this, a check should be made using a plane frame analysis of the section to compare the distortional stiffness of the prototype cell with that of a cell of the same geometry but having identical top and bottom slabs whose combined stiffnesses equal those of the prototype.) It is also assumed that the cross-section distorts as in Fig. 10.7(b) without horizontal sideways deflection of the slabs.
9. The values of the influence lines at 'beams' to the side of the 'beam' considered decay in geometric progression. In other words, the values of the points on each influence line in Charts C of Figs 10.2–10.4 decrease in geometric progression away from the peak values. This assumption is discussed later.

Charts A and B in Figs 10.2–10.4 enable the user to derive the peak value of the influence line for a particular 'beam' from the non-dimensional parameters f, r and c. Chart C is used to derive the rest of the influence line through the peak value. The influence value J at any point across an influence line for a 'beam' is the fraction of the total moment on the deck, due to a load above the point, that is carried by the 'beam'. Alternatively it can be thought of as the moment (or deflection) in the 'beam' expressed as a fraction of the total moment (or deflection) that the 'beam' would experience if it carried the load by itself without load distribution to the rest of the deck. In fact the shape of the various influence lines is not uniquely defined by the peak value, and variations from the assumed geometric progression occur as shown in Fig. 10.6. However, since the sum of the values of J at all the points on a line must sum to unity, a small overestimate of J at one point is compensated for by an underestimate elsewhere. It is unlikely that design loads will be so distributed across a section that significant errors ensue. The reason the geometric decay has been assumed is that it provides a reasonable fit to the folded plate output and at the same time provides a quick method of calculation of the influence line values.

The Charts C have been called 'influence lines' as it is generally more convenient to use them as such. More strictly they are distributions of moment or deflection for the first harmonic of a line load on the relevant 'beam'. As long as the 'beams' are not very close together (so that distribution of harmonics higher than the first is not significant), the reciprocal theorem can be applied with little error to moments in the 'beams' in addition to deflections as is strictly correct. Consequently, Charts C can be used either as distributions of moment and deflection for a load on a particular 'beam' or as influence lines.

Most design loads that are critical for longitudinal moments consist of a distributed load or a number of related point loads near midspan. The distribution of moments and deflections across the deck approximate closely to the distribution of the first harmonic as assumed in the charts. But, under the action of badly distributed loads such as a single point load or concentration of loads at one end, the distribution of moments can be worse than that of the first harmonic charts. To compensate for possible error on this count, it is generally worth taking the precaution of Morice and Little [5] of arbitrarily increasing the calculated design moments by 10%.

Since the critical design load for shear force often consists of a concentration of load near one end, the distribution is worse than that of the first harmonic charts. Correction can be made as outlined in Section 12.3.3, but in general it is simpler to assume that only loads between the quarter span points are distributed while those near the ends are not distributed.

Charts C only give the values of each influence line at the 'beam' positions. The shape of the line between 'beams' depends on the rotational stiffness parameter r. If r is small as it is for the decks of Fig. 10.1(a) and (b), the deck cross-section distorts in a smooth curve. In contrast, if r is large, as for the decks of Fig. 10.1(c) and (d), the distortion and influence lines are 'stepped'. The rotation at the 'beams' can be estimated from Charts D which give the ratio of the 'beam' rotation ϕ to the average rotation $\bar{\phi}$ calculated from the relative deflection of the 'beams' on each side, as shown in Fig. 10.7. In the case of the edge 'beam', $\bar{\phi}$ is calculated from the relative deflection of the edge and penultimate 'beams'.

10.4 APPLICATION OF CHARTS TO SLAB DECK

Figure 10.8(a) and (b) show a solid slab deck supporting an abnormal heavy vehicle. The charts of Figs 10.2–10.4 will be used to derive influence lines for moments at points across the cross-section, from which will be derived the moments due to the applied load.

The deck is first notionally split up into identical 'beams' as shown in

Fig. 10.8 Influence lines for slab deck: (a) elevation; (b) cross-section; (c) cross-section divided into 'beams'; and (d) influence lines for 'beams'.

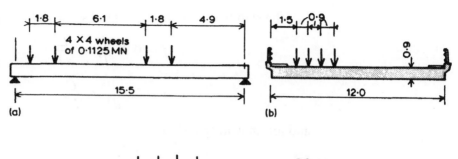

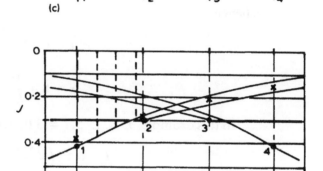

Fig. 10.8(c). The optimum width of 'beam' for isotropic slabs is of the order of 1/5 to 1/4 of the span.

For an isotropic slab, orthotropic slab or beam-and-slab deck the non-dimensional parameters are

$$f = 0.12 \, \frac{i}{l^3} \times \frac{L^4}{I} \tag{10.1}$$

$$r = 5 \, \frac{G}{E} \times \frac{l}{i} \times \frac{C}{L^2} \tag{10.2}$$

where

L = span
l = 'beam' spacing
i = transverse moment of inertia of slab per unit length
I = moment of inertia of 'beam'
C = torsion constant of 'beam' + $l \times$ (transverse torsion constant of slab per unit length).

If the slab has depth d and $\nu = 0.2$ so that $G/E = 0.42$,

$$f = 0.01 \frac{d^3}{l^3} \times \frac{L^4}{I} \tag{10.3}$$

$$r = 25 \frac{l}{d^3} \times \frac{C}{L^2} \tag{10.4}$$

and for an isotropic slab

$$I = \frac{ld^3}{12}$$

$$C = \frac{ld^3}{6} + l\left(\frac{d^3}{6}\right) = \frac{ld^3}{3}$$

hence

$$f = 0.12 \frac{L^4}{l^4} \tag{10.5}$$

$$r = 8.4 \frac{l^2}{L^2}. \tag{10.6}$$

In this example, $L = 15.5$, $l = 3.0$, hence

$$f = 0.12 \left(\frac{15.5}{3.0}\right)^4 = 86$$

$$r = 8.4 \left(\frac{3.0}{15.5}\right)^2 = 0.31.$$

Figure 10.9 demonstrates how Fig. 10.2 is used to derive the influence line for the edge 'beam' with $f = 86$ and $r = 0.31$. The procedure is:

1. Determine the point on Fig. 10.2, Chart A where the vertical through $f = 86$ cuts the contour of $r = 0.31$. The ordinate J_1, here

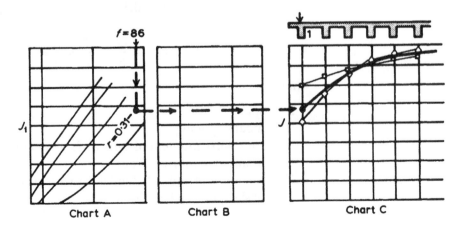

Fig. 10.9 Application of Fig. 10.2 for $f = 86$ and $r = 0.31$.

equal to 0.33, is the value of the peak ordinate under the edge beam of the influence line in Chart C. Chart B is ignored for slab and beam-and-slab decks.

2. In Fig. 10.2, Chart C, interpolate the line of the influence line passing through $J_1 = 0.33$ and which lies between the lines shown for J_1 greater and less, as shown in Fig. 10.9. Hence we find the values of J at all the beams.

$$\downarrow$$
$$J \quad 0.33 \quad 0.22 \quad 0.15 \quad 0.10 \quad \Sigma = 0.80. \tag{10.7}$$

On a wide deck, the influence line decays to zero on the far side and the sum of all the values of J should be unity. If in addition the values of J decay in geometric progression,

$$\frac{J_n}{J_{n-1}} = (1 - J_1) \tag{10.8}$$

it will be found that the ratio of adjacent figures in equation (10.7) is equal to $(1 - 0.33)$. In a narrow deck such as this, the sum of all the influence values should also be unity and not 0.80 as indicated in equation (10.7). Consequently the figures in equation (10.7) should be corrected by scaling up by 1/0.80 to give a sum of unity

$$\downarrow$$
$$J \quad 0.41 \quad 0.28 \quad 0.19 \quad 0.12 \quad \Sigma = 1.0. \tag{10.9}$$

These are the values of the influence line for edge beam 1 plotted in Fig. 10.8(d).

The same procedure is followed to obtain the influence line for beam next to the edge, but using Fig. 10.3. For $f = 86$ and $r = 0.31$ we find $J_2 = 0.20$. By following the influence line in Fig. 10.3, Chart C, through $J_2 = 0.20$ we obtain values of J at other beams.

$$\downarrow$$
$$J \quad 0.20 \quad 0.20 \quad 0.15 \quad 0.11 \quad \Sigma = 0.66. \tag{10.10}$$

Although J_1 must be derived from Fig. 10.3, Chart C (or with the reciprocal theorem from J_2 on edge beam line) the values of J to the right of J_2 can be found using geometric reduction factor

$$\frac{J_n}{J_{n-1}} = \frac{(1 - J_1 - J_2)}{(1 - J_1)}. \tag{10.11}$$

This factor should differ very little from that of equation (10.8). As before, the values of J in equation (10.10) appropriate to an infinitely wide deck must be scaled up to give a sum of unity, giving

$$\downarrow$$
$$J \quad 0.30 \quad 0.30 \quad 0.23 \quad 0.17 \quad \Sigma = 1.0. \tag{10.12}$$

This influence line for 'beam' 2 is also plotted in Fig. 10.8(d). Also shown are the lines for 'beams' 3 and 4 which are mirror images of lines for 'beams' 1 and 2.

Although this deck has no internal 'beams' for which Fig. 10.4 is appropriate, it is worth stating here the geometric reduction factor for influence lines in Chart C to have geometric decay and sum of unity. The factor is

$$\frac{J_n}{J_{n-1}} = \frac{(1-\hat{J})}{(1-\hat{J})}. \tag{10.13}$$

This factor should also differ very little from that of equation (10.8).

The total moment on the deck at the cross-section under the nearest axle to midspan due to one line of wheels is

$$M = 0.1125(0.46 + 1.16 + 3.80 + 2.78)$$
$$= 0.9225 \text{ MNm}.$$

The lines of wheels coincide with points on the influence line for 'beam' 1 at which $J = 0.41, 0.37, 0.33, 0.30$. Hence, the moment at 'beam' 1 due to four lines of wheels is

$$M_1 = 0.9225(0.41 + 0.37 + 0.33 + 0.30) = 1.30 \text{ MNm}.$$

Similarly,

$$M_2 = 0.9225(0.30 + 0.30 + 0.30 + 0.30) = 1.11 \text{ MNm}$$
$$M_3 = 0.9225(0.17 + 0.19 + 0.21 + 0.23) = 0.74 \text{ MNm}$$
$$M_4 = 0.9225(0.12 + 0.14 + 0.16 + 0.18) = 0.55 \text{ MNm}.$$

For design purposes, these moments are increased by 10%.

Figure 10.8(d) also shows, by crosses, the influence line for 'beam' 1 obtained from the load distribution charts of Morice and Little [5]. It can be seen that at 'beam' 1 the value obtained from the charts of this book is an overestimate by 7%; however the calculated maximum live load moment in 'beam' 1 is overestimated by only 3%. When the live load moment is combined with that due to dead load, this error becomes negligible. This is because the charts were derived on the assumption that the deck is very wide. By considering only a narrow width and assuming the decay rate of influence values is the same as for a wide deck, the transverse rotation of the deck and the loading of the edge under an eccentric load are overestimated. Generally, the error has very little effect on calculated maximum design moments as it only affects narrow decks with high transverse stiffness whose load distribution is good.

Fig. 10.10 Influence lines for beam-and-slab deck: (a) elevation; (b) cross-section; and (c) influence lines for 'beams'.

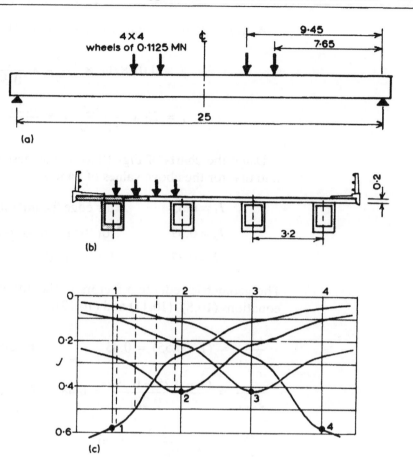

10.5 APPLICATION OF CHARTS TO BEAM-AND-SLAB DECK

10.5.1 Spaced box beams

Figure 10.10 shows a beam-and-slab deck constructed of prestressed box beams supporting a reinforced concrete slab. The deck is supporting an abnormal heavy vehicle of four axles, each of four wheels, located at midspan between the edge beam and the next beam to the edge. The reinforced concrete has a modular ratio $m = 0.8$ relative to the prestressed concrete.

For analysis, the deck is divided into four identical 'beams' at $l = 3.2$ m spacing. For each 'beam'

$$I = 0.30 \, \text{m}^4$$

$$C = 0.26 + 3.2 \left(\frac{0.8 \times 0.2^3}{6} \right) = 0.26 \, \text{m}^4.$$

With $L = 25.0$ we obtain from equations (10.3) and (10.4)

$$f = 0.01 \times 0.8 \times \frac{0.2^3}{3.2^2} \times \frac{25^4}{0.30} = 2.5$$

$$r = 25 \times \frac{3.2}{0.8 \times 0.2^3} \times \frac{0.26}{25^2} = 5.2.$$

Using the charts of Figs 10.2–10.4 as described in Section 10.4, we find that for the above values of f and r

$J_1 = 0.55$ for edge 'beam' influence line

$J_2 = 0.39$ for 'beam' next to edge influence line

$\hat{J} = 0.37$ for internal 'beam' influence line.

The geometric reduction factors in the influence lines are then, from equations (10.8), (10.11) and (10.13),

$$\frac{J_n}{J_{n-1}} = (1 - 0.55) = 0.45 \quad \text{for edge 'beam' influence line}$$

$$\frac{J_n}{J_{n-1}} = \frac{(1 - 0.25 - 0.39)}{(1 - 0.25)} = 0.48 \quad \text{for 'beam' next to edge influence line}$$

$$\frac{J_n}{J_{n-1}} = \frac{(1 - 0.37)}{(1 + 0.37)} = 0.46 \quad \text{for internal 'beam' influence line.}$$

Using these factors or working directly from Charts C of Figs 10.2–10.4 we obtain influence lines for edge and next to edge (and internal 'beams' not here relevant)

$$\downarrow$$
$$J \; 0.55 \; 0.25 \; 0.11 \; 0.05 \; \Sigma = 0.96 \qquad (10.14)$$

$$\downarrow$$
$$J \; 0.25 \; 0.39 \; 0.19 \; 0.09 \; \Sigma = 0.92 \qquad (10.15)$$

$$\downarrow$$
$$(J \; 0.08 \; 0.17 \; 0.37 \; 0.17 ...) \qquad (10.16)$$

The sums of the influence values in equations (10.14) and (10.15) are close to unity because the load distribution characteristics of the deck are not very good, and the deck is nearly wide enough for the moments in 'beams' 1 and 2 to be little influenced by loads or additional beams on the far side. On factoring up equations (10.14) and (10.15) to give sums of unity we obtain

$$J \;\; \overset{\downarrow}{0.58} \;\; 0.26 \;\; 0.11 \;\; 0.05 \;\; \Sigma = 1.0 \tag{10.17}$$

$$J \;\; 0.27 \;\; \overset{\downarrow}{0.42} \;\; 0.21 \;\; 0.10 \;\; \Sigma = 1.0. \tag{10.18}$$

These influence lines are plotted in Fig. 10.10(c) together with the mirror image lines appropriate to 'beams' 3 and 4.

The box beams all have high torsion stiffness so that $r = 5.2$. It can be seen in Charts D of Figs 10.2–10.4 that for such a value of r, $\phi/\bar{\phi} = 0.5$ for all 'beams', showing that the rotation of the 'beams' is only about half of the average rotation in that region of deck. Consequently, the influence lines of Fig. 10.10(c) are drawn with gradients at the beam positions equal to 0.5 of the gradient of the line drawn through the 'beam' points on each side. For example, using values in equation (10.17), we find that the rotation of 'beam' 2 is

$$\phi_2 = 0.5 \frac{(0.58 - 0.11)}{2l}. \tag{10.19}$$

In general, it is not necessary to calculate the beam rotations as ϕ can be drawn approximately by eye equal to the relevant fraction of $\bar{\phi}$.

For the abnormal heavy vehicle loading, the midspan moment per longitudinal line of wheel is

$$M = 0.1125(7.65 + 9.45) = 1.924 \text{ MNm}. \tag{10.20}$$

Since the lines of wheels coincide with values on influence lines for 'beams' 1, 2, 3 and 4 of

'beam' 1	0.57	0.49	0.35	0.27
'beam' 2	0.28	0.31	0.38	0.42
'beam' 3	0.11	0.13	0.17	0.20
'beam' 4	0.05	0.07	0.09	0.11

the total moments in the three 'beams' are

$$M_1 = 1.924(0.57 + 0.49 + 0.35 + 0.27) = 3.23 \text{ MNm}$$
$$M_2 = 1.924(0.28 + 0.31 + 0.38 + 0.42) = 2.67 \text{ MNm}$$
$$M_3 = 1.924(0.11 + 0.13 + 0.17 + 0.20) = 1.17 \text{ MNm}$$
$$M_4 = 1.924(0.05 + 0.07 + 0.09 + 0.11) = 0.62 \text{ MNm}.$$

For design, these should all be arbitrarily increased by 10%.

A grillage analysis of the same deck subjected to the same load case gave moments in the four beams of 3.16 MNm, 2.52 MNm, 1.33 MNm and 0.69 MNm, respectively.

10.5.2 Contiguous beam deck

A contiguous beam deck can be analysed by the charts or grillage with each analysis 'beam' representing more than one physical beam. Figure 10.5 gives an example where each analysis 'beam' represents two physical beams. This simplification does not affect the calculated load distribution unless the summed torsional stiffness of the analysis 'beam' is large enough to cause the cross-section to distort in steps as in Fig. 10.6(b). This condition can be considered satisfied if

$$r < 0.5$$

or

$$l < 0.02 \frac{d^3 L^2}{C}. \tag{10.21}$$

This limitation of beam spacing applies to slabs and beam-and-slab decks for analysis by chart or grillage.

10.6 APPLICATION OF CHARTS TO CELLULAR DECK

Figure 10.11 shows a cellular deck with thin webs and top and bottom slabs of different thicknesses. The ratio of stiffnesses of the top and bottom slabs is

$$\left(\frac{0.2}{0.15}\right)^3 = 2.4$$

which is less than the limiting value for these charts stated in Section 10.3.

For a cellular deck, the non-dimensional parameters are

$$f = 0.12 \frac{i}{l^3} \times \frac{L^4}{I} \tag{10.22}$$

$$r = 6 \frac{i_w}{i} \times \frac{l}{h} \tag{10.23}$$

$$c = 0.1 \frac{G}{E} \times \frac{I_s}{I} \times \frac{L^2}{l^2} \tag{10.24}$$

Application of charts to cellular deck 217

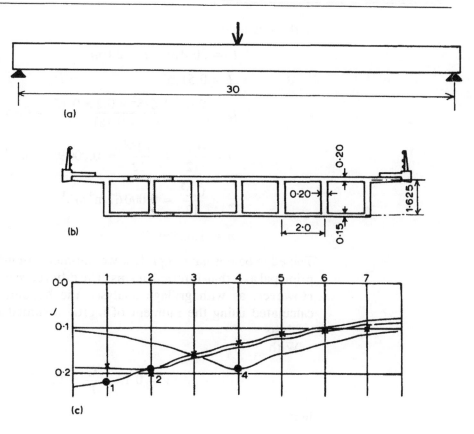

Fig. 10.11 Influence lines for cellular deck: (a) elevation; (b) cross-section; and (c) influence lines for 'beams'.

where

L = span

l = 'beam' spacing = web spacing

I = moment of inertia of 'beam'

I_s = moment of inertia of slabs between webs about neutral axis of deck for one 'beam'

i = sum of individual moments of inertia, per unit length, of top and bottom slabs

i_w = moment of inertia, per unit length, of the web for bending out of its place

h = height of web between midplanes of slabs.

In this example,

$$L = 30.0 \text{ m} \quad l = 2.0 \text{ m}$$
$$I = 0.51 \text{ m}^4$$
$$I_s = \frac{2.0 \times 1.625^2 \times 0.2 \times 0.15}{(0.2 + 0.15)} = 0.45 \text{ m}^4$$
$$i = \frac{0.2^3}{12} + \frac{0.15^3}{12} = 0.00095 \text{ m}^4 \text{ m}^{-1}$$
$$i_w = \frac{0.2^3}{12} = 0.00067 \text{ m}^4 \text{ m}^{-1}$$
$$h = 1.625 \text{ m}.$$

The edge beams have slightly lower moments of inertia about the deck principal axis than internal beams. The difference here is not significant. However, as with grillage analysis, the bending stresses should be calculated using the moment of inertia assumed in load distribution analysis.

With

$$v = 0.15 \quad \frac{G}{E} = 0.435$$

hence

$$f = 0.12 \times \frac{0.00095}{2.0^3} \times \frac{30^4}{0.51} = 23$$

$$r = 6 \times \frac{0.00067}{0.00095} \times \frac{2.0}{1.625} = 5.2$$

$$c = 0.1 \times 0.435 \times \frac{0.45}{0.51} \times \frac{30^2}{2^2} = 8.6.$$

The application of the charts of Figs 10.2–10.4 using all three parameters is demonstrated in Fig. 10.12, which reproduces Fig. 10.2.

1. Determine the point on Fig. 10.2, Chart A, where the vertical through $f = 23$ cuts the contour of $r = 5.2$.
2. Move across Chart A to the right edge and then move along the interpolated contour line on Chart B until the vertical through $c = 8.6$ is reached. The ordinate of J_1, here 0.15, is the value of the peak ordinate of the influence line at the beam in Chart C. The reduction of influence value J_1 on traversing the contour in Chart B represents the load distribution due to cellular torsion.

Fig. 10.12 Application of Fig. 10.2 for $f = 23$, $r = 5.2$ and $c = 8.6$.

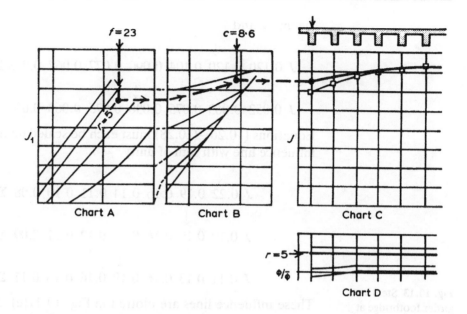

3. On Chart C, interpolate the line of the influence line passing through $J_1 = 0.15$ which lies between the lines shown for J_1, above and below on the chart. In this case J_1 is less than the lowest line and hence the decay away from the 'beam' must be calculated with the geometric reduction factor of equation (10.8). Here

$$\frac{J_n}{J_{n-1}} = (1 - 0.15) = 0.85.$$

Hence the influence line for edge 'beam' is

$$\downarrow$$
$$J\ \ 0.15\ \ 0.128\ \ 0.108\ \ 0.092\ \ 0.078\ \ 0.066\ \ 0.057\ \ \Sigma = 0.679 \quad (10.26)$$

Applying Figs 10.3 and 10.4 in a similar manner we find for

$$\text{'beam' next to edge} \quad J_2 = 0.12$$
$$\text{internal 'beam'} \quad \hat{J} = 0.09.$$

From equations (10.11) and (10.13), the appropriate geometric reduction factors are:

$$\text{'beam' next to edge} \quad \frac{J_n}{J_{n-1}} = \frac{(1 - 0.12 - 0.12)}{(1 - 0.12)} = 0.86$$

$$\text{internal 'beam'} \quad \frac{J_n}{J_{n-1}} = \frac{(1 - 0.09)}{(1 + 0.09)} = 0.83$$

with which we calculate the 'beam' values on the influence lines for

'beams' 2 and 4

$$\downarrow$$
$$J\ 0.120\ 0.120\ 0.104\ 0.090\ 0.077\ 0.067\ 0.058\ \Sigma = 0.636 \quad (10.27)$$
$$\downarrow$$
$$J\ 0.052\ 0.063\ 0.075\ 0.090\ 0.075\ 0.063\ 0.052\ \Sigma = 0.47. \quad (10.28)$$

Equations (10.26)–(10.28) must each be scaled up to give the corrected influence line with sum of unity.

$$\downarrow$$
$$J\ 0.22\ 0.19\ 0.16\ 0.14\ 0.11\ 0.10\ 0.08\ \Sigma = 1.0$$
$$\downarrow$$
$$J\ 0.19\ 0.19\ 0.16\ 0.14\ 0.12\ 0.11\ 0.09\ \Sigma = 1.0$$
$$\downarrow$$
$$J\ 0.11\ 0.13\ 0.16\ 0.19\ 0.16\ 0.13\ 0.11\ \Sigma = 1.0.$$

These influence lines are plotted in Fig. 10.11(c). Since the factor r is equal to 5.2, Charts D of Figs 10.2–10.4 indicate that $\phi/\bar{\phi} = 0.5$ as in the last example. Hence influence lines are stepped, as in the last example,

Fig. 10.13 Steel box-girder footbridge at Chichester, England; designed and built by Butterley Engineering Ltd. Photograph E.C. Hambly.

with rotations at 'beams' approximately equal to half the average rotation between 'beams' on two sides.

The influence lines are used for the calculation of moments in the same manner as demonstrated in the previous two examples.

The deck of this example has proportions which differ only slightly from those of a perspex model reported by Sawko [10]. The model, of span 36.8 in, had six cells of inside dimensions 2×2 in with all webs and slabs 0.25 in thick. For these dimensions, $f = 31$, $r = 3$, $c = 8.6$; it will be found in the charts that these non-dimensional parameters give the same influence lines and deflection distributions as the example above. Consequently, the lines of Fig. 10.11(c) can also be considered as chart predictions for the deflection profiles of the model. Sawko reported the predictions of finite element analyses; the deflection profile for a load above the web next to the edge is shown in Fig. 10.11(c) by crosses. It is evident that the chart prediction is very close.

REFERENCES

1. Jaeger, L.G. and Bakht, B. (1989) *Bridge Analysis by Microcomputer*, McGraw-Hill, New York.
2. Pucher, A. (1964) *Influence Surfaces of Elastic Plates*, Springer-Verlag, Vienna and New York.
3. Rusch, H. and Hergenroder, A. (1961) *Influence Surfaces for Moments in Skew Slabs*, Technological University, Munich. Translated from German by C.R. Amerongen, Linden, Cement and Concrete Association.
4. Balas, J. and Hanuska, A. (1964) *Influence Surfaces of Skew Plates*, Vydaratelstvo Slovenskej Akademic Vied, Bratislava.
5. Morice, P.B. and Little, G. (1956) *Right Bridge Decks Subjected to Abnormal Loads*, Cement and Concrete Association, London.
6. Rowe, R.E. (1962) *Concrete Bridge Design*, CR Books, London.
7. Cusens, A.R. and Pama, R.P. (1975) *Bridge Deck Analysis*, John Wiley, London.
8. Department of the Environment (1970) Ministry of Transport Technical Memorandum: Shear Key Decks, Annexe to Technical Memorandum (Bridges), No BE 23.
9. Hambly, E.C. (1976) *Bridge Deck Behaviour*, Chapman and Hall, London, 1st edn.
10. Sawko, F. (1968) 'Recent developments in the analysis of steel bridges using electronic computers', British Constructional Steelwork Association Conference on Steel Bridges, London, June, pp. 39–48.

11 Temperature and prestress loading

11.1 INTRODUCTION
This chapter describes the actions of temperature loading and prestress on bridge decks, and shows how these loads can be simulated in the analytical methods outlined in previous chapters. Although temperature loading and prestress are described below one after the other, it is important to emphasize the very different principles of these two types of loading.

11.2 TEMPERATURE STRAINS AND STRESSES IN SIMPLY SUPPORTED SPAN
The rise of temperature in an element of material causes the element to expand if it is unrestrained as shown in Fig. 11.1(a). Alternatively, if the element is prevented from expanding as shown in Fig. 11.1(b) the rise in temperature causes an increase in stress which depends on Young's modulus of the material. Either the increase in strain or the increase in stress can be taken as the starting point for the calculation of the distribution of temperature stresses in the structure. For a simply supported deck with linear variation in temperature between top and bottom surfaces, the simplest way to calculate the flexure is to work from the temperature-induced free strain, which causes flexure without stress, as shown in Fig. 11.1(c). However, for the more general bridge problems with complicated deck geometry and non-linear temperature distributions, it is simplest to start from the assumption that the deck experiences temperature changes which set up stresses while the deck is rigidly restrained throughout, as shown in Fig. 1.11(d), and then calculate the effects of removing the theoretical restraints.

Figure 11.2(a) shows an element of deck with non-linear temperature distribution varying from +21°C at the top surface through 0 to +7°C at the soffit. (The temperature distribution is derived from a heat flow

Fig. 11.1 Effects of temperature distribution and restraint conditions on movements and stresses in elements of bridge deck.

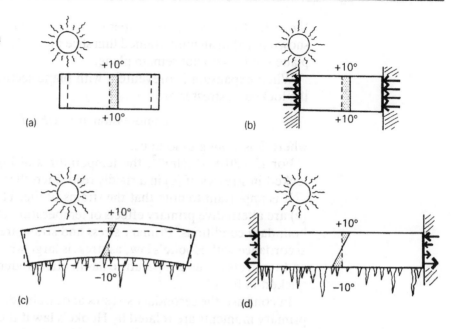

Fig. 11.2 Temperature effects in slab deck: (a) temperature distribution; (b) strains in unrestrained slice; (c) stresses in restrained slice; (d) = (c); and (e) compression P, and (f) restrained moment M_T and residual.

calculation as described in references [1]–[3], or from the appropriate code of practice.)

If the coefficient of thermal expansion is α, the unrestrained thermal strains are

$$\text{expansion } \varepsilon = \alpha \Delta T. \tag{11.1}$$

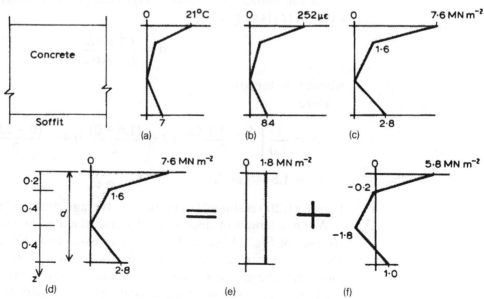

If $\alpha = 12 \times 10^{-6}\,°\mathrm{C}^{-1}$, the temperatures of Fig. 11.2(a) cause the strains shown in (b) in an unrestrained thin slice of the deck. With such strains, plane sections do not remain plane.

When expansion is prevented, with plane sections held plane, then the locked-in stresses are

$$\text{compression } \sigma = \alpha E \Delta T \qquad (11.2)$$

where E is Young's modulus.

For $E = 30\,000\,\mathrm{MN m^{-2}}$, the temperatures of Fig. 11.2(a) cause the locked-in stresses of (c) in a rigidly restrained slice of deck.

It is important to note that the strains of Fig. 11.2(b) and stresses of (c) are alternative primary effects of temperature change dependent on boundary conditions. Primary stress and strain are not proportional in accordance with Hooke's law, as stress is large when strain is kept small and vice versa. The temperature has in effect induced a 'lack of fit' or a 'jacked-in force'.

In contrast, the secondary stresses and strains due to redistribution of primary moments are related by Hooke's law if the material is elastic.

The locked-in stress distribution of Fig. 11.2(c), shown again in (d), can be thought of as composed of two parts which affect the structure in different ways:

1. an average compression stress shown in Fig. 11.2(e) causing resultant force on cross-section of P;
2. a non-linear stress distribution shown in Fig. 11.2(f) which comprises a moment M_T and residual stresses with no net resultants.

For the particular stresses of Fig. 11.2(d), the average compression stress of (e) is simply

$$\bar{\sigma} = \int_0^d \frac{\sigma b\, dz}{\text{Area}} \qquad (11.3)$$

where $b = $ breadth.

Here

$$\bar{\sigma} = \frac{1}{1.0}\left[\frac{(7.6+1.6)}{2}\,0.2 + \frac{(1.6+0)}{2}\,0.4 + \frac{(0+2.8)}{2}\,0.4\right]$$

$$= 1.8\,\mathrm{MN\,m^{-2}}.$$

Figure 11.2(f) is simply (d) minus this average stress in (e).

When a length of deck is rigidly restrained as in Fig. 11.3(a), the stresses of Fig. 11.2(d) act on every cross-section. On internal cross-sections the stresses on two sides balance, so that it is only at the end faces that these stresses must be balanced by the externally applied restraining forces. When longitudinal expansion restraint is removed, as

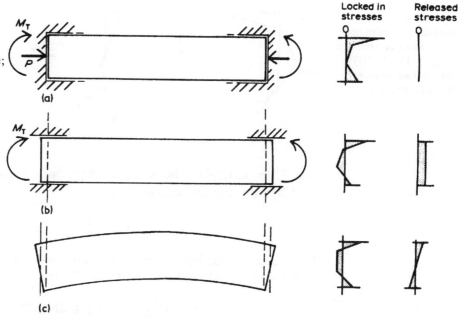

Fig. 11.3 Progressive release of temperature forces: (a) fully restrained; (b) compression release; and (c) flexure release.

in Fig. 11.3(b), the deck expands and the compression stresses of Fig. 11.2(e) are relaxed. Only the moment and residual stresses of Fig. 11.2(f) remain. When the moment restraint is removed as in Fig. 11.3(c) the deck flexes. In effect, the restraining moments have been cancelled by equal and opposite relaxing moments which cause flexure. The magnitudes of the restraining (and opposite relaxing) moments are equal to the moment of the stress diagram of Fig. 11.2(f) (about any point).

$$M = \int_0^d \sigma b z \, dz. \qquad (11.4)$$

Here

$$M = \frac{5.8}{2} \times 0.193 \times 0.064 - \frac{0.2}{2} \times 0.007 \times 0.198$$

$$- \frac{(0.2 + 1.8)}{2} \times 0.4 \times 0.45 - \frac{1.8}{2} \times 0.26 \times 0.69$$

$$+ \frac{1.0}{2} \times 0.14 \times 0.95$$

$$= -0.24 \, \text{MN m m}^{-1} \, \text{width}. \qquad (11.5)$$

The stresses and strains caused by the relaxing moment are related by Hooke's law and exhibit the elastic behaviour described in Chapters 2

Fig. 11.4 Temperature moment and residual stresses: (a) restrained moment M_T and residual (same as Fig. 11.2(f)); (b) relaxing moment M_T; and (c) residual.

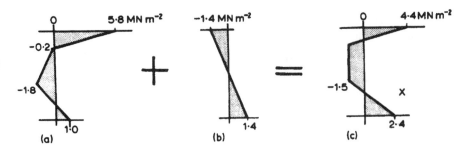

and 3. In particular, the stress distribution is linear as shown in Fig. 11.4(b) with stress related to M by

$$\sigma = \frac{M}{Z}. \qquad (11.6)$$

For a rectangular beam $Z = bd^2/6$, and hence the stress relaxation at surfaces in Fig. 11.4(b) due to $M = -0.24$ is

$$\sigma = -\frac{0.24}{1^2/6} = -1.4\,\text{MN}\,\text{m}^{-2}.$$

When the relaxing stress distribution of (b) is added to the restrained stresses of (a), we obtain the residual stress distribution of (c). This has no net compression or moment, as can be checked by reapplying equations (11.3) and (11.4).

The residual stress diagram Fig. 11.4(c) shows the final distribution of temperature-induced stresses on a cross-section of a simply supported span. The maximum stresses are large, even though the deck is simply supported, because the temperature distribution is so non-linear. It is only when the temperature distribution is linear that its induced stresses are linear in (a) so that when relaxed by an equal and opposite linear stress distribution of (b), no residual stresses (c) remain on the section.

At the end faces of the deck, the residual temperature stresses are not resisted by external forces. Consequently, these forces redistribute by local elastic distortion over a length of deck approximately equal to the depth of the section. This redistribution is accompanied by local high longitudinal shear forces which transfer the residual compression forces near the top and bottom faces to the opposed residual tension in the middle. In this example, the residual compression force below level X in Fig. 11.4(c) is

$$\frac{2.4}{2} \times 0.25 = 0.3\,\text{MN}\,\text{m}^{-1}\,\text{width}.$$

Thus the longitudinal shear force at level X near the end of the deck is also $0.3\,\text{MN}\,\text{m}^{-1}$ width.

In the above example it is assumed that the deck has the same breadth at all levels. If the deck is made up of beams with breadth b varying with depth, the variation in b is included in equations (11.3) and (11.4).

Further worked examples of calculations of temperature stresses are included in references [3]–[6].

The preceding discussion has assumed that the bridge material is able to transmit compression and tension forces at all levels within the cross-section. This is not the case for reinforced concrete which is assumed for the purposes of design to have cracks in regions subject to tension. These cracks are generally advantageous with respect to temperature effects because they can accommodate the thermal strains without developing stresses. References [3] and [4] explain the calculation procedure which follows the principles of this chapter.

11.3 TEMPERATURE STRESSES IN A CONTINUOUS DECK

The relaxation of temperature stresses in a continuous deck is only slightly more complicated than in a simply supported span. Figure 11.5(a) shows a three-span deck which is fully restrained with a locked-in non-linear temperature distribution.

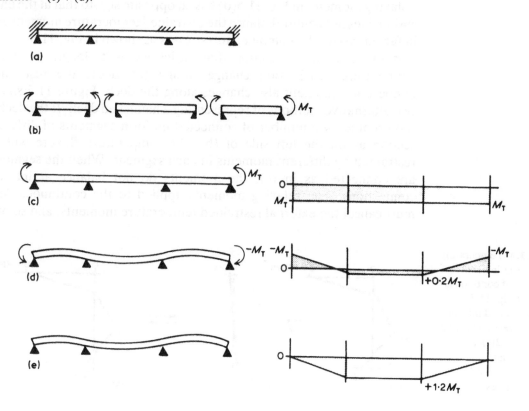

Fig. 11.5 Release of temperature moments in continuous deck: (a) restrained deck; (b) restrained moments in each span; (c) restrained moments in connected spans; (d) relaxing flexural moments; and (e) final moments = (c) + (d).

The compression force due to average temperature compression stress of Fig. 11.2(e) is relaxed by letting all the spans expand freely in the longitudinal direction. Figure 11.5(b) shows the restrained temperature moments on each span similar to Fig. 11.3. Since the moments on the two sides of the internal supports balance, it is possible to join the spans together as in (c) without restraints at internal supports and without any relaxation taking place. In other words, only moments at the ends of the deck are needed to restrain a continuous deck against temperature flexure. Relaxation of these end restraining moments is achieved by superposing equal and opposite relaxing moments as shown in Fig. 11.5(d). In contrast to a simply supported span which hogs with uniform moment when subjected to end relaxing moments, the continuous deck distributes the relaxing moments in accordance with the continuous beam theory of Chapter 2 as shown in Fig. 11.5(d). Consequently, superposition of the stresses from the restrained temperature moments of (c), which are uniform along the deck, and the varying relaxing moments of (d) give different stress distributions at different points along the deck.

At the ends of the end spans, the relaxing moment is equal and opposite to the restrained temperature moment, and the stress distribution is similar to Fig. 11.4(c). However at the next support, the 'relaxing moment' in Fig. 11.5(d) has an opposite sign to that at the ends and so does not counterbalance the restrained temperature moment and in fact increases the combined top stresses as shown in Fig. 11.6.

If the deck has a section that varies along its length or if the temperature distribution changes along the deck, the restrained temperature moment also changes along the deck. Figure 11.7 shows two alternative methods of representing a haunched deck. If the deck is represented by a number of connected uniform segments of different section as on the left side of (b), the temperature flexure will be restrained by different moments in each segment. When the segments are connected as in (c), the moments do not balance at internal connections. The relaxing moments applied to the continuous deck must cancel the external restrained temperature moments, and so out-

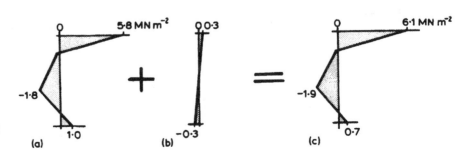

Fig. 11.6 Temperature stresses at internal support of continuous deck of Fig. 11.5: (a) restrained moment M_T and residual (same as Fig. 11.2(f)); (b) relaxing moment = $0.2\,M_T$; and (c) total stresses.

Fig. 11.7 Release of temperature moments in haunched deck: (a) deck elevation; (b) restrained temperature forces on each span; (c) restrained temperature forces on connected spans; and (d) relaxing flexural forces.

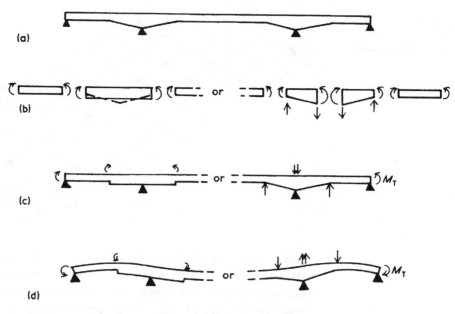

of-balance moments must be applied at changes of section of continuous beam as in (c). Alternatively, if a segment is tapered, the restrained temperature moments are different on its two ends. To maintain equilibrium, opposed vertical forces must be applied to the two ends of the segment, as on the right of (b). (This shear force is statically equivalent to a uniformly distributed moment applied along the taper.) When the segments are connected, the restrained temperature moments now balance at the connections but the vertical restraining forces remain. As a consequence, the cancelling relaxing forces applied to the continuous beam for distribution include vertical forces at the ends of tapers as well as moments at the end supports.

Wide bridges experience stresses due to the thermal expansion of an element being resisted transversely as well as longitudinally. The behaviour is the same in the two directions. However if Poisson's ratio is significant, it may be necessary to investigate the interaction of stresses and flexure in the two directions. This is best done with a finite element analysis in which the interaction is taken account of automatically.

11.4 GRILLAGE ANALYSIS OF TEMPERATURE MOMENTS

The analysis of distribution of temperature moments in a two-dimensional deck by grillage is the same in principle as that for

Fig. 11.8 Relaxing flexural moments applied to grillage.

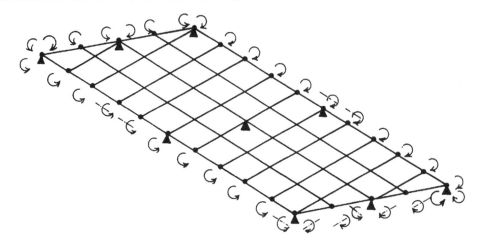

continuous beam in Section 11.3. Figure 11.8 shows a grillage for a two-span skew deck. The restrained temperature moments and cancelling flexural moments on every elemental beam are similar to those on the elements of Fig. 11.5(b). When the elements are connected, only the out-of-balance moments (and forces) remain applied to the joints. In this example it is assumed that the slab deck is uniform so that the moments on the element beams on the two sides of the internal joints cancel. At the edges, the temperature-induced stresses of Fig. 11.4(a) must be resisted by equal and opposite external forces while flexure is restrained. When these moments are cancelled by relaxing moments, they are only applied to the edges of the deck as in Fig. 11.8. The relaxing moments distribute non-uniformly throughout the deck as in the continuous beam in Fig. 11.5(b), and the precise distribution must be found from the grillage analysis. The final stresses at any point are simply the superposition of the stresses from the grillage output moments and the restrained temperature stresses of Fig. 11.4(a).

In wide multicell decks with few diaphragms, the top and bottom slabs at points far from a diaphragm can expand and contract with little resistance from the deck except the out-of-plane flexure of the webs. A three-dimensional analysis is necessary if a detailed study of temperature stresses is to be made of such a deck or of any other complex structure. Most three-dimensional structural analysis programs have the facility to input directly different temperatures in members without the user having to calculate equivalent loads or lack of fit.

11.5 DIFFERENTIAL CREEP AND SHRINKAGE

The physical action of differential creep or shrinkage in a structure is similar to that of temperature. If a thin slice of deck is unrestrained, differential creep and shrinkage will cause non-uniform strains in the slice in the same way as temperature in Fig. 11.2(b). The only difference is that the creep and shrinkage strain diagram is usually stepped. Also, if the slice is restrained rigidly, the differential creep and shrinkage usually induce a tensile stress distribution (similar to temperature compression Fig. 11.2(c)) with tensile stresses equal to the effective modulus × the free creep and shrinkage strains. The distribution of secondary moments is calculated in the same manner as for temperature moments.

11.6 PRESTRESS AXIAL COMPRESSION

A prestressing cable in a concrete bridge deck subjects the concrete to three different systems of loading, illustrated in Fig. 11.9:

1. axial compression in concrete due to compression force applied to concrete by anchorage of cable or friction forces, shown in Fig. 11.9(a);
2. moments due to eccentricity of resultant compression force from the neutral axis of concrete section, shown in (b);
3. vertical loads on concrete due to reaction of curved cable, shown in (c).

This and subsequent sections give a brief description of how the concrete structure behaves under these load systems, and then demonstrate how the loads are simulated for application to the analytical models described in preceding chapters. The general behaviour of prestressed concrete and its design is described much more thoroughly in references [7] and [8].

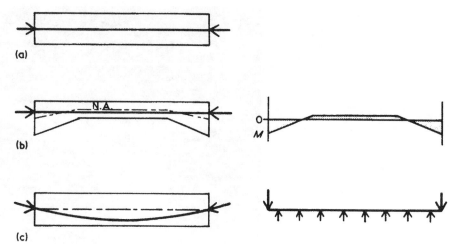

Fig. 11.9 Forces on concrete due to prestress: (a) axial compression; (b) moments due to eccentricity of compression; and (c) vertical (or horizontal) forces due to cable curvature.

Fig. 11.10 Multicellular concrete box girder viaduct at Staples Corner, London, England; designed, tested and built by Taylor Woodrow Construction. Photographs bottom and middle courtesy of Taylor Woodrow Construction; top E.C. Hambly.

The prestress compression force on a long beam subjects cross-sections from end to end to a uniform stress equal to the force divided by the section area. The distribution of stress is only complicated in the region of the anchorage and ends of flanges where local shear lag action is significant. The distribution of prestress compression stress in a wide deck can also be simple if the prestress anchorages are evenly distributed across the section, or if the density of prestress varies linearly from one side to the other. In the latter case the deck is bent in plan, and complications arise only if the supports are rigid against the small sideways deflections. However, if the deck has a complicated shape in plan, if the prestress cables are curved in plan, or if one part is stressed more (or before) the rest, some form of three-dimensional analysis may be necessary. This could be a space frame (described in Section 11.10) or a finite element analysis (described in Chapter 13).

11.7 PRESTRESS MOMENTS DUE TO CABLE ECCENTRICITY

Figure 11.11(a) shows a continuous deck with constant cross-section subjected to prestress by a straight cable with force P at eccentricity e from the neutral axis. Because the concrete only comes into contact and reacts with the cable at the end anchorages, the only forces applied to the concrete are compression P and hogging moment Pe at anchorages. Since the neutral axis is straight, the compression force P simply subjects the whole length of deck to uniform compression, which does not affect continuous beam analysis. The end moments make the deck flex, but because it is held down by internal supports, the moments distribute non-uniformly as shown in Fig. 11.11(c). This distribution is found from continuous beam analysis with beam subjected to end moments as in (b).

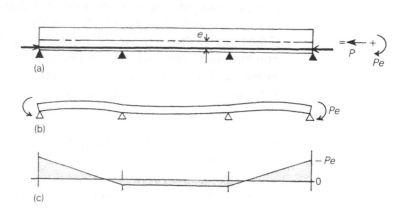

Fig. 11.11 Prestress moments in three-span deck due to eccentricity of prestress: (a) elevation; (b) moments applied to concrete; and (c) moment diagram.

Fig. 11.12 Prestress moments due to eccentricity in deck of varying section:
(a) elevation;
(b) compression on restrained elements;
(c) moments due to eccentricity of compression on restrained elements;
(d) moments and forces applied in continuous beam analysis; and
(e) diagram of distributed moments.

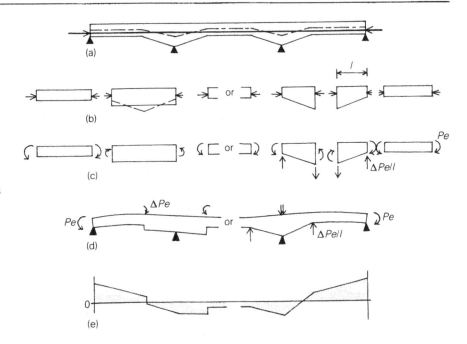

When the deck has a varying section as in Fig. 11.12(a), the changing eccentricity of the axial force P subjects the concrete to varying applied moment along its length. The effect can be visualized by considering the beam (or grillage) restrained at the support positions and at the changes of section. The fixed end moments Pe are applied to each end of each section, and a continuous beam or grillage analysis is then used to determine the redistribution on release and removal of the restraints. At internal changes of section the fixed end moments Pe on each side largely cancel, and only the difference ΔPe need be applied. In tapers, as on the right of Fig. 11.12, Pe is the same on both sides of any section and so cancels out. However, the concrete along the taper is subjected to moment ΔPe which must be applied uniformly along the taper, or, as is much simpler, this uniformly distributed applied moment is represented by statically equivalent vertical loads $\Delta Pe/l$ on the two ends of the taper as shown in Fig. 11.12(c) and (d). Changes ΔPe of prestress moment along the deck due to decrease of P by friction can also be represented in the beam or grillage analysis by opposed vertical loads $\Delta Pe/l$ at the two ends of the frictional change.

11.8 PRESTRESS MOMENTS DUE TO CABLE CURVATURE

A curved prestress cable presses against the concrete on the inside of the curve and thus subjects the deck to vertical loads. The cable profile is

Fig. 11.13 Prestress loading due to cable curvature: (a) forces on concrete and (b) loading for continuous beam or grillage analysis.

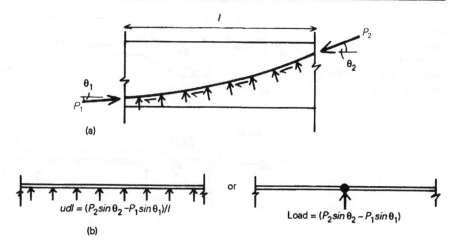

usually chosen so that these vertical loads on the concrete largely cancel out the loads and bending stresses in concrete due to self weight. Since the vertical prestress forces press on the concrete in the same way as live or dead loads, their effects can be investigated with continuous beam or grillage in precisely the same way.

Figure 11.13 shows the forces acting on a short length of concrete due to a curved prestressing tendon. If the inclination of the cable changes from θ_1 to θ_2 while the prestress force changes from P_1 to P_2 due to friction

$$\text{vertical force} = P_2 \sin \theta_2 - P_1 \sin \theta_1 \quad (11.7)$$

If the cable curve has a parabolic shape, the load is uniformly distributed with

$$\text{u.d.l.} = (P_2 \sin \theta_2 - P_1 \sin \theta_1)\frac{1}{l}. \quad (11.8)$$

Strictly, the vertical loads due to prestress should be applied to the continuous beam or grillage analysis as distributed loads with intensity varying with curvature. But if the beam is notionally split up into short elements as in a grillage, the prestress vertical loads can be applied as point loads at the joints with θ_1 and θ_2 in equation (11.7) being the cable gradients midway along the elements to each side.

Figure 11.14 summarizes the loading that has to be applied to a continuous beam or grillage to simulate the effects of anchorage eccentricity, varying section and cable curvature. Working from left to right the loads are:

1. hogging moment Pe_1 simulating moment on concrete at anchorage due to eccentricity e_1 of anchorage below the neutral axis;

Fig. 11.14 Example of prestress loading for continuous beam analysis: (a) cable profile and (b) loading.

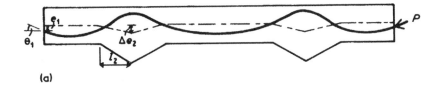

(a)

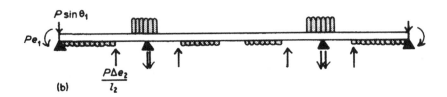

(b)

2. vertical force $P\sin\theta_1$ simulating vertical component of force on anchorage due to inclination θ_1 of cable;
3. upwards distributed load along length of deck that cable is curving upwards;
4. vertical forces $P\Delta e_2/l$ at each end of haunch to simulate bending loads on deck due to change in level Δe_2 of the neutral axis along haunch;
5. downwards distributed load above support along length of deck that cable is curving downwards.

It should be noted that the sum of all vertical loads on the deck due to prestress should be zero.

11.9 PRESTRESS ANALYSIS BY FLEXIBILITY COEFFICIENTS

The analysis of prestress moments in Sections 11.7 and 11.8 is orientated towards a grillage analysis in which the structure is in effect held rigidly at the joints, subjected to the vertical loads and fixed end moments due to prestress on the elements, and released to redistribute the moments. An alternative approach can be used in which the structure is first considered as simply supported spans subjected to the moments due to prestress, and then the reactant moments are found that are necessary to make the segmented structure join together. This is the flexibility coefficient method described in Section 2.3.5.

Figure 11.15 shows a slice of deck. The prestress cable crosses the two faces of the slice with tensions P_1 and P_2 and eccentricities e_1 and e_2. When the deck is simply supported, the compression P in the concrete (shown in Fig. 11.15) at any section is coincident and opposite to the cable tension. The free moment in the concrete at that point is simply

$$M = Pe \qquad (11.9)$$

Fig. 11.15 Prestress forces on free element of deck: (a) forces and (b) equivalent moments.

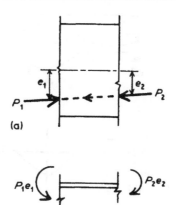

where P is the prestress force at the section and e is the eccentricity of the prestress cable from the neutral axis. If the cable is curved and the level of the neutral axis moves up and down as in Fig. 11.16(a), e is still the distance between these two lines. (b) shows the free moment diagram along the spans. It is similar in shape to the plot of the cable off the neutral axis, except that the moment is reduced away from the anchorage due to the drop in P by friction. Under the action of these free moments, the various spans will deflect and, since they are at present considered as separate spans, relative rotations occur between adjacent ends of spans on each side of the internal supports. At support 1 the relative rotation is given by equation (2.12):

$$\delta_{1W} = \int \frac{m_1 m_w \, ds}{EI} \qquad (11.10)$$

where m_w is the free moment Pe due to applied prestress and m_1 is the moment due to the unit reactant moment at support 1, shown in Fig. 11.16(d). To calculate the product integral (equation 11.10) it is necessary to plot Pe/EI as in (c) by dividing the moment diagram of (b) by local EI. The product integral can then be computed by multiplying (c) by m_1 in (d), thus giving δ_{1W}. The relative rotation δ_{2W} at support 2 is found in a similar manner. By following the procedure outlined in Section 2.3.5, δ_{12} (and δ_{21}) is found, but with change of EI considered. Equations of the form of equation (2.11) are then obtained to give reactant moments X_1 and X_2 at the supports in Fig. 11.16(e). These moments are added to the free moments of (b) to give the total moments in the continuous concrete structure as shown in (f).

It should be noted that for simply supported spans, prestress is simply an interaction between cable and concrete and there are no external reactions. When the spans are made continuous the reactant moments X_1, X_2, etc. induce reactions at the support. Often the free moments are

Fig. 11.16 Flexibility analysis of prestress moments in continuous beam: (a) cable profile; (b) free moment $M = Pe$ diagram; (c) free M/EI diagram; (d) moments for unit release at 1; (e) reactant, parasitic (or secondary) moments; and (f) total prestress = (b) + (e).

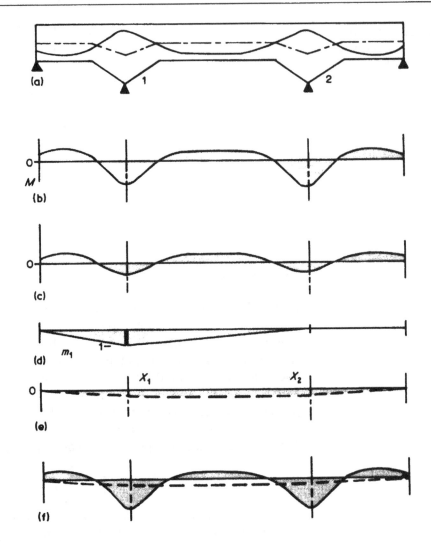

referred to as 'primary prestress' and the reactant moments as 'secondary' or 'parasitic' moments. This distinction is convenient for continuous beams, but it has little meaning in wide two-dimensional decks which are skew, tapered or curved since the lateral redistribution of moments due to the deck itself is as significant as longitudinal redistribution due to continuity at supports.

11.10 PRESTRESS APPLIED DIRECTLY TO SPACE FRAME

The discussion on prestress effects in Sections 11.6 to 11.9 largely relates to analyses by continuous beam and grillage. Many of the complications in deriving equivalent loads can be avoided if a plane frame or space

Prestress applied directly to space frame

frame analysis is used with the prestress applied to the model as an internal load, in the same manner as it is applied to the bridge structure. There are also advantages during the interpretation of results because they can be related directly to the design without having to worry whether they are 'primary' or 'secondary' effects.

The prestress internal load can be applied to the space frame, either as forces, or as a shortening of bars representing the prestress tendons. The latter method is demonstrated below with two examples. The first example relates to a pretensioned beam in which the bars representing strands are attributed the strands' actual cross-section area and stiffness, and the deformation of the model reduces the prestress. This realistically simulates the loss of pretension at transfer from the pretensioning beds. The second example relates to a post-tensioned structure in which the structure deforms without reducing the prestress. In this case the prestress is represented by bars with Young's modulus factored down (and subjected to prestrains factored up) so that the distortions of the model do not reduce the prestress.

11.10.1 Pretensioned beam

Figure 11.17 illustrates a plane frame model of half of a pretensioned prestressed concrete beam with composite reinforced concrete slab. The slab is shown dotted because it does not form part of the model at the time of transfer of prestress. At a later stage of analysis it is added to analyse the redistribution of stresses due to temperature, differential shrinkage and creep.

Each stiffness in the model in Fig. 11.17 is based on the part of the section it represents. Longitudinal members A represent the concrete of the bottom flange, as shown in Fig. 11.17(b). Members B represent the web longitudinally while C represent the web vertically for transfer of longitudinal shear. These members have shear areas modified as explained for the cruciform space frame in Chapter 7, using equation

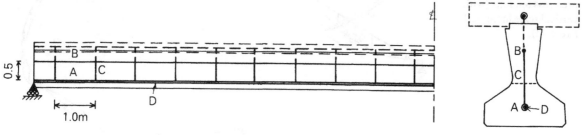

Fig. 11.17 Plane frame model of half length of composite prestressed concrete beam with reinforced concrete slab (shown dotted): (a) space frame; (b) section.

(7.6). For members A and B dimensions $a = 1.0$ m and $b = 0.5$ m, while members C have shear areas calculated with $a = 0.5$ m and $b = 1.0$ m. Members A, B and C have Youngs modulus $E = 31\,000$ MPa. Members D represent the prestress strands which in this case have their centroid collinear with the centroid of the bottom flange A. These members are given the area of all the prestress strand with $E = 200\,000$ MPa. The area of members D decreases towards the end to represent the debonding of strands.

The pretensioning is applied at the works by stressing the tendons to 75% of their characteristic strength which is $0.75 \times 1670 = 1250$ MPa. This means that the tendons are subjected to an initial strain of $(1250/200\,000) = 0.006\,250$. If this tensile strain is applied as a load case to members D of the model in Fig. 11.17(a) it reproduces the effect of the transfer of the prestress. This behaviour may be understood by initially fixing the joints at the ends of the tendons in the model. The initial strain then sets up the initial prestress force which is balanced by the tensions out of the ends of the strands at the fixed joints. When these joints are released (i.e. the jacking load is slackened off) the prestress force is then transferred to compression along the concrete. The shortening of the concrete leads to a reduction of the prestress, which is the elastic loss at transfer. Subsequent creep strains can be modelled by modifying Young's modulus of the concrete.

11.10.2 Post-tensioned deck with draped tendons

Figure 11.18 illustrates the space frame model for the prestressed concrete box-girder deck in Fig. 9.10, with the vertical scale exaggerated fivefold to show the prestress profile. Longitudinal

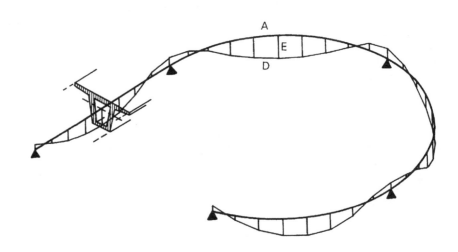

Fig. 11.18 Space frame model of one span of a curved concrete box girder with draped prestress tendons (with vertical scale increased five-fold).

members A in the space frame represent the box-section at the level of its centroid. Members D represent the prestress tendons which have the draped profile as shown. They are held in place relative to the box neutral axis by the outriggers E which are stiff. Longitudinal members A are given the section properties of the box girder as for a continuous beam, while members D are given the cross-section of the tendons. The prestress is applied as in the previous example by applying a tensile strain to members D so as to generate the appropriate tensile force. But in order to reproduce a post-tensioning force which is not influenced by shortening of the structure, the Young's modulus of the tendons is reduced tenfold (say) and the tensioning strain in members D increased by the same factor. Variations in prestress along the length due to friction losses can be represented by variations in the applied strains. When the tension forces are applied to members D they subject the girder A to vertical load at each joint where there is a change in direction of the prestress. This directly simulates the loading due to curvature of the prestress tendons.

Figure 11.19(a) illustrates the bending moment diagrams derived from the curved model due to a prestress force of 14 MN in members D and due to a uniformly distributed load of 0.1 MN/m on members A. Figure 11.19(b) shows the torsions in the structure due to the same loads. It is evident that the prestress induces a different distribution of torque to the distributed load. Although the torque from the prestress is smaller it extends over a greater length and so causes a larger twisting rotation.

The space frame model can be made more detailed, if desired, to reproduce the complexities of the cross-section. For example, separate longitudinal members can be used to represent the top flange, webs and bottom flange. Then the effects of temperature differentials can be investigated directly by subjecting the different longitudinal members to different thermal strains.

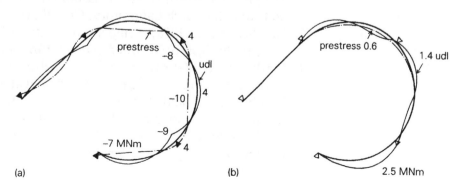

Fig. 11.19 (a) Moments along curved spans of Fig. 11.18 due to distributed load and draped pre-stress tendons and (b) torques.

Fig. 11.20 Multispan cellular concrete deck of Greta Bridge, Keswick, England; concrete cast in-place and post-tensioned. Designed by Scott Wilson Kirkpatrick & Partners. Photograph courtesy Scott Wilson Kirkpatrick & Partners.

REFERENCES
1. Priestley, M.N.J. (1972) Thermal gradients in bridges – some design considerations, *New Zealand Engineering*, 15 July, pp. 228–33.
2. Emerson, M. (1977) *Temperature Differences in Bridges: Basis of Design Requirements*, TRRL Laboratory Report 765, Transport and Road Research Laboratory, Crowthorne.
3. Hambly, E.C. (1978) Temperature distributions and stresses in concrete bridges, *Structural Engineer*, **56A**, No. 5, pp. 143–8.
4. Clark, L.A. (1983) *Concrete Bridge Design to BS5400*, Construction Press, Longmans, London, with supplement (1985).
5. Hambly, E.C. and Nicholson, B.A. (1991) *Prestressed Beam Integral Bridges*, Prestressed Concrete Association, Leicester.
6. Steel Construction Institute (1989) 'Design Guide for Continuous Composite Bridges: 1 Compact Sections' by D.C. Iles, SCI Publication 065.

7. Libby, J.R. and Perkins, N.D. (1976) *Modern Prestressed Concrete Highway Bridge Superstructures*, Van Nostrand Reinhold, New York.
8. Harris, J.D. and Smith, I.C. (1963) *Basic Design and Construction in Prestressed Concrete*, Chatto and Windus, London.

12 Harmonic analysis and folded plate theory

12.1 INTRODUCTION

The grillage and finite element methods described in other chapters analyse complicated bridge decks by considering them as assemblages of simple structural elements, for each of which simplified force–deflection behaviour is assumed. Numerous discontinuities of structure are introduced to make solution possible. One method of analysis which does not require the structure to be cut up in the same way uses harmonic analysis and either 'orthotropic' or 'folded' plate theory. Exact solutions are possible in the sense that the mathematical model can be made rigorously to satisfy the stress–strain assumptions for an elastic continuum. However the methods, as described below, can be used only for prismatic decks whose cross-section is the same from abutment to abutment. Figure 12.1 shows examples of cross-sections of common types of deck which do have such sections. (a) and (b) can be analysed by orthotropic plate theory, while (c), (d) and (e) can be analysed by folded plate theory.

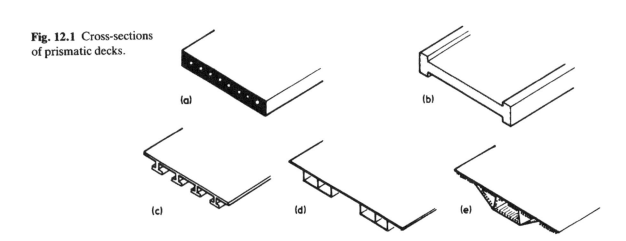

Fig. 12.1 Cross-sections of prismatic decks.

Fig. 12.2 Breakdown of load into sinusoidal harmonic components.

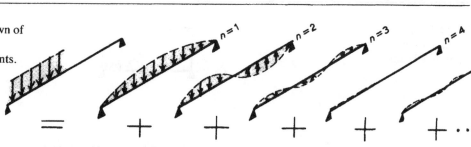

In harmonic analysis, the load is broken down into a number of components, each consisting of a distributed line load parallel to the structure and with intensity varying as a pure sine wave as in Fig. 12.2. Under the action of each sine wave component, every longitudinal slab or web of the structure deflects and twists in a pure sine wave. Since every differential of a sine wave is a sine or cosine function, the equilibrium equations, which can be thought of as differentials of deflections, can also be expressed as a number of sine or cosine functions without differential symbols. These equations can be solved as conventional simultaneous equations.

This chapter explains how a load or other function can be represented as a number of harmonic components. It is then shown how the load distribution characteristics of a deck are markedly different for low and high harmonics so that the severity of different types of load can be anticipated from their harmonic composition without detailed analysis of the structure. Finally, the basic principles of folded plate theory are summarized.

12.2 HARMONIC COMPONENTS OF LOAD, MOMENT, ETC.

The general theory of harmonic, or Fourier, analysis is described in detail in many mathematics text books and handbooks, including reference [1]. This section describes only a particular application of the theory to deck analysis. Figure 12.3(a) shows a simply supported beam of span L with coordinate origin at the left support. The beam supports a load. It is possible to define a generalized infinite series of harmonic components for the loading, and then calculate the magnitudes of the harmonics relevant to the boundary conditions of zero moment and zero deflection at support positions. However, it is simpler to satisfy the boundary conditions using physical reasoning. The beam and load can be considered as part of an infinitely long beam with repetitive reversed loading as in (b). Automatically there are points of zero moment and deflection at points of contraflexure spaced regularly L apart. The repetitive reversed loading W, which is a function of x, can be

246   Harmonic analysis and folded plate theory

Fig. 12.3 (a) Simple span as part of (b) infinite beam.

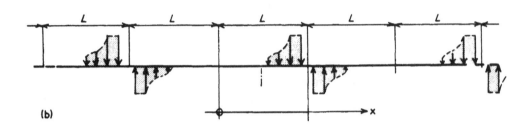

represented by an infinite series of sine functions of the form

$$W(x) = b_1 \sin \frac{\pi x}{L} + b_2 \sin \frac{2\pi x}{L} + \cdots + b_n \sin \frac{n\pi x}{L} + \cdots. \quad (12.1)$$

There are no cosine terms in this case because on differentiating twice and four times for moment and deflection, cosine functions do not satisfy the boundary conditions of no moment or deflection at $x = 0$ and L.

The values of the coefficients b are given by

$$b_n = 2 \times \left[\text{ average value of } W(x) \sin \frac{n\pi x}{L} \text{ between } x = 0 \text{ and } x = L \right]. \quad (12.2)$$

For example, for the case of the uniformly distributed load shown in Fig.

Fig. 12.4 Uniformly distributed load on simple span.

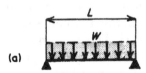

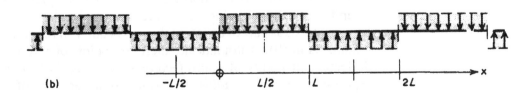

12.4(a),

$$b_n = 2 \times \frac{1}{L} \int_0^L \left(\frac{W}{L}\right) \sin \frac{n\pi x}{L}$$

$$= \begin{cases} 2 \times \dfrac{1}{L} \times \dfrac{W}{L} \times \dfrac{2L}{n\pi} & \text{for } n = 1, 3, 5, \ldots \\ 0 & \text{for } n = 2, 4, 6, \ldots \end{cases} \quad (12.3)$$

All the coefficients for even number harmonics in the above example are zero because the uniformly distributed load is symmetric about the deck centre line, whereas even number harmonic functions are all antisymmetric. It is possible to represent any load case, such as the point load in Fig. 12.5, as a combination of a symmetric component shown in (b), and an antisymmetric component shown in (c). The harmonic representation of the symmetric component is composed entirely of odd number harmonic functions, while the antisymmetric component is represented entirely by even number harmonics.

Thus for load, moment or deflection:

$$\begin{aligned} &\text{Symmetric component harmonics} \quad n = 1, 3, 5, \ldots \\ &\text{Antisymmetric component harmonics } n = 2, 4, 6, \ldots \end{aligned} \quad (12.4)$$

The shear force diagrams for the load case of Fig. 12.5(a) and its symmetric and antisymmetric components in (b) and (c) are shown in Fig. 12.6(a), (b) and (c). These should also be thought of as parts of infinitely long repetitive diagrams. It should be noted that the

Fig. 12.5 Symmetric and antisymmetric components of a load: (a) total; (b) symmetric; and (c) antisymmetric.

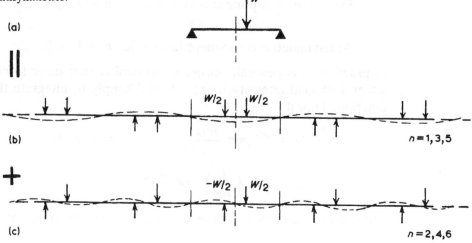

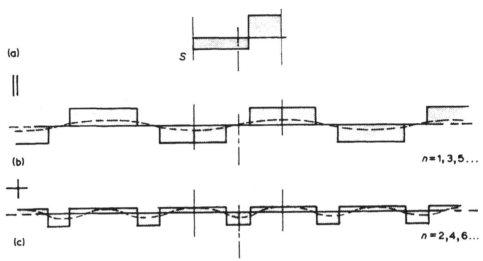

Fig. 12.6 Shear force diagram for loads of Fig. 12.5: (a) total; (b) antisymmetric; and (c) symmetric.

symmetric load produces an antisymmetric shear force diagram and vice versa.

The general expression for the shear force diagram can be written

$$S(x) = a_1 \cos \frac{\pi x}{L} + a_2 \cos \frac{2\pi x}{L} + \cdots + a_n \cos \frac{n\pi x}{L} + \cdots \quad (12.5)$$

where

$$a_n = 2 \times \left[\text{average value of } S(x) \cos \frac{n\pi x}{L} \text{ between } x = 0 \text{ and } x = L \right]. \quad (12.6)$$

It will be found for shear force and slope that:

Symmetric component harmonics $\quad n = 2, 4, 6, \ldots$

$$(12.7)$$

Antisymmetric component harmonics $n = 1, 3, 5, \ldots$

In practice, it is generally easier to remember that shear force is the integral of load intensity times -1, and simply to integrate the load function. Thus if

$$W = \Sigma b_n \sin \frac{n\pi x}{L} \qquad n = 1, 2, 3, \ldots$$

$$S = \Sigma b_n \left(\frac{L}{n\pi} \right) \cos \frac{n\pi x}{L} \quad (12.8a)$$

and similarly, for bending moment, slope and deflection we obtain from repeated integration

$$M = \Sigma b_n \left(\frac{L}{n\pi}\right)^2 \sin \frac{n\pi x}{L}$$

$$EI \frac{dw}{dx} = \Sigma b_n \left(\frac{L}{n\pi}\right)^3 \cos \frac{n\pi x}{L} \qquad (12.8b)$$

$$EIw = \Sigma b_n \left(\frac{L}{n\pi}\right)^4 \sin \frac{n\pi x}{L}.$$

It is useful to remember that, for any harmonic, the coefficient of the load, bending moment and deflection are proportional in the ratio

$$1 : \left(\frac{L}{n\pi}\right)^2 : \left(\frac{L}{n\pi}\right)^4 \times \frac{1}{EI}.$$

The amplitudes of harmonic components for typical design loads are tabulated in Figs A.2 and A.3. Column 1 of Fig. A.3 shows the total load function with its integrals for shear force, etc., derived from simple beam theory. Column 2 gives the magnitude of the first harmonic. Column 3 gives the sum of all higher harmonics, which is simply the difference between columns 1 and 2. Column 4 gives the amplitude of any other harmonic n.

12.3 CHARACTERISTICS OF LOW AND HIGH HARMONICS

12.3.1 Distribution of low harmonics

An understanding of the harmonic composition of load can be particularly useful in obtaining insight into the physical behaviour of beam-and-slab or cellular bridge decks. The basic form of such decks, with longitudinal stiffness large by comparison with transverse stiffness, accentuates the different response to low harmonic and high harmonic loads. For example, consider the bridge deck shown in Fig. 12.7. Column 1 shows a load acting on one web together with the distributed bending moments and deflected form. Columns 2 and 3 show the first and the third harmonics for each function in Column 1.

When beam 1 deflects under a load, some of this load is transferred to beams 2 and 3 by the vertical shear associated with out-of-plane flexure and torsion of the slab and some by the in-plane shear of the slab. Although the slab is thin by comparison with the depth of the beam, its span between beams is short, giving a stiffness (for spanning between beams) comparable to that of the beams (spanning between abutments). Thus when beam 1 deflects, a significant proportion of the first harmonic of moment is transferred through the slab to beam 2. This is shown in Fig. 12.7, column 2, in which it can be seen that beam 1 is carrying about 60% of the first harmonic bending moment while beams 2 and 3 are carrying about 25 and 15% respectively. Precise amounts depend on the structure's form and dimensions. It should also be noted that, for a deck composed of identical beams, the distribution of first harmonics is identical for deflections and bending moments because for any harmonic component, bending moment is proportional to deflection.

12.3.2 Concentration of high harmonics

Under the action of the third harmonic of loading, the bridge deck in Fig. 12.7 is, in effect, simply supported between points of contraflexure of the third harmonic sine wave. Its span is effectively one third of the first harmonic span and, consequently, its longitudinal bending stiffness becomes 27 times that of the first harmonic. On the other hand, transverse bending of the slab is as before, with the slab spanning

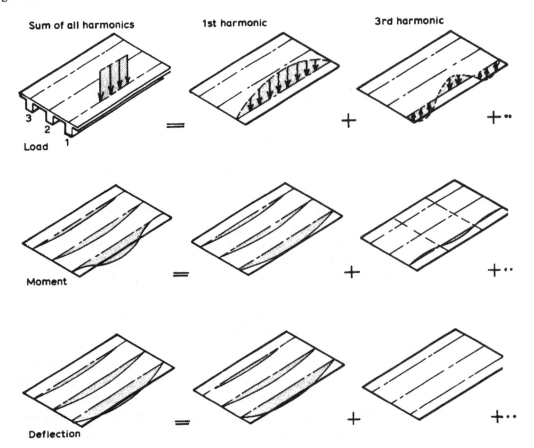

Fig. 12.7 Distribution of load, moment and deflection across a bridge deck.

between beams. Consequently, when the loaded beam deflects under the action of the third harmonic of the load, nearly all the load is carried longitudinally by the very stiff beam, and very little transverse distribution takes place. A deck distributing the first harmonic in the manner of Fig. 12.7, column 2, would retain approximately 90% of the third harmonic in the loaded beam, distributing only 10% to the adjacent beam as shown in column 3.

As a working hypothesis, it is often convenient to assume that the first harmonic of the load is distributed transversely as shown in Fig. 12.7, column 2 while higher harmonics remain concentrated in the loaded beam. This is strictly only applicable to simply supported bridge decks with 'beams', as defined in Chapter 10, at centres greater than about one tenth of the span. However, it is often possible to analyse midspan sections of continuous bridge decks by considering the parts of the deck between points of contraflexure as being simply supported. In the general analysis of bridge decks which are continuous or have beams at close centres, it is necessary to analyse the higher harmonics, up to the level where distribution is considered negligible.

Another way of looking at the structural property of non-distribution of higher harmonics is to consider it as an example of St Venant's principle, (see Section 1.3). Figure 12.8 shows part of a slab subject to a high harmonic load. In the region of the wave shown, the slab must deflect downwards under the downward half wave and upwards along the upward half wave. However, at a point further than a half wavelength away, the effects of the downward and upward loads virtually cancel out. In other words, local variation in the load only affects a width similar to the length over which the variation occurs, while over a larger area the large number of upward and downward forces are self-cancelling. Consequently, if an abnormal vehicle is standing on a bridge deck, precise details of the distribution of the point loads have little effect on the magnitude of bending in the beams, whose scale is much

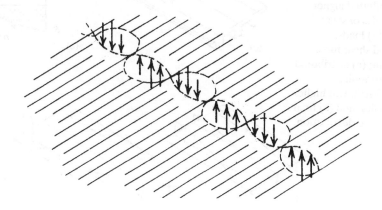

Fig. 12.8 Localized effect of high harmonic load.

larger than distances between the wheels. But these distances do affect the local slab bending moments because the span of the slab is of the same scale. For this reason it is valid to superpose separate analyses for overall distributed bending moments and for local slab moments under the wheels. However, it is important that the correct loads be applied in the analysis of overall deck moments, and that correct boundary conditions are assumed in the local bending analysis.

When a bridge deck is stiffened longitudinally by a few beams or webs at wide centres, it is important that the load should not simply be statically distributed between beams. The transverse flexural behaviour of the slab can be compared to that of a continuous beam with rotationally stiff elastic supports. In the load distribution analysis, the true load can be replaced by the fixed end moments and shear forces, and it should not be replaced by statically distributed loads on the supports. Strictly, the fixed edge moments and shear forces should be obtained from the local load analysis with assumed fixed edge boundary conditions.

In contrast, loads can usually be statically redistributed locally along beams since there are no sudden variations in stiffness in that direction. Similarly, local statical redistribution of loads on a thick slab deck has little effect on the distributed load behaviour.

12.3.3 Example of separation of effects of low and high harmonics

Figure 12.9 shows how the distribution of shear force near an abutment can be found by superposing the distributed first harmonic and the undistributed higher harmonics. Figure 12.9(a) shows the point loads of

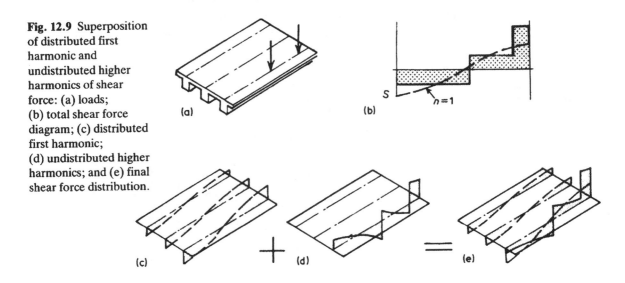

Fig. 12.9 Superposition of distributed first harmonic and undistributed higher harmonics of shear force: (a) loads; (b) total shear force diagram; (c) distributed first harmonic; (d) undistributed higher harmonics; and (e) final shear force distribution.

two wheels of a vehicle near the abutment. (b) shows the shear force diagram for the bridge as a whole. The total shear force, i.e. the sum of all harmonics, which is shown shaded, was obtained using simple beam theory. The first harmonic component (shown dotted) was found using the equations of Section 12.2. As explained previously, the first harmonic distributes significantly and this is shown in (c). The precise distribution relevant to the deck construction is obtained from the charts of Chapter 10. The sum of the higher harmonics is the difference in (b) between the total shear force and the dotted first harmonic. These higher harmonics are assumed to remain undistributed and are applied solely to the loaded beam as in (d). The final shear force distribution in (e) is found by recombining the distributed first harmonic in (c) with the undistributed higher harmonics in (d).

It is evident in Fig. 12.9(e) that the final distribution of shear force differs significantly from the distribution of the first harmonic (c) obtained from charts. This is because the higher harmonics form a significant part of the total shear force when the load is near one abutment. Appendix A Fig. A.3 gives first harmonic and higher harmonic components for several design loads. It can be seen that, in general, the first harmonic component of moment, slope or deflection approximates closely to the total function, and it is only for shear force near the load that the discrepancy can be very significant.

12.4 HARMONIC ANALYSIS OF PLANE DECKS

Several publications listed in the references [1–9] outline the application of harmonic theory to the analysis of various types of deck in greater detail than is possible here. To demonstrate here how harmonic analysis can be used, the artificially simplified deck of Fig. 12.10 is analysed below.

The beam-and-plank deck of Fig. 12.10(a) is constructed of three box beams supporting a running surface of transversely spanning planks. To simplify the problem it is assumed that the beams have such high torsional stiffnesses that they do not twist, so that the only deflections of the structure are vertical deflections w_1, w_2 and w_3 of the beams as shown in (b). The length of deck shown represents the distance between points of contraflexure of the nth harmonic. With the origin at the left end the deflections of beams 1 and 2 at any point are

$$w_1 \sin \alpha x \quad \text{and} \quad w_2 \sin \alpha x$$

where

$$\alpha = \frac{n\pi}{L}. \tag{12.9}$$

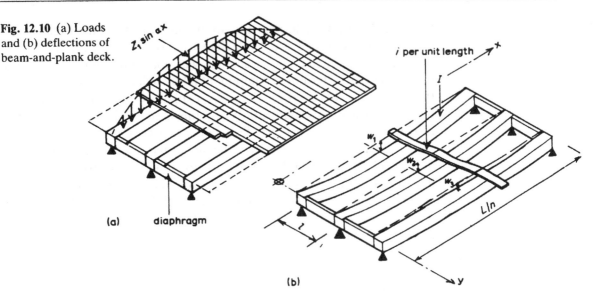

Fig. 12.10 (a) Loads and (b) deflections of beam-and-plank deck.

The vertical shear force in the transverse spanning planks per unit length of deck is given by equation (2.8) and is

$$s_{12} = \frac{6Ei}{l^2}\left[\frac{2}{l}(w_1 \sin \alpha x - w_2 \sin \alpha x)\right]$$

$$= \frac{12Ei}{l^3}(w_1 - w_2)\sin \alpha x \qquad (12.10)$$

where i = moment of inertia of planks per unit length of deck.

If $Z_1 \sin \alpha x$ is the nth harmonic of the applied load above beam 1, the net vertical load on beam 1 at any point is

$$Z_1 \sin \alpha x - s_{12} = \left[Z_1 - \frac{12Ei}{l^3}(w_1 - w_2)\right]\sin \alpha x. \qquad (12.11)$$

It is shown in equations (12.8) that if a beam of inertia I is subjected to load $b_n \sin \alpha x$, its deflection is $(b_n/EI\alpha^4)\sin \alpha x$. Consequently, if the load on beam 1 is given by equation (12.11) and the deflection is $w_1 \sin \alpha x$, then

$$w_1 \sin \alpha x = \frac{1}{EI\alpha^4}\left[Z_1 - \frac{12Ei}{l^3}(w_1 - w_2)\right]\sin \alpha x$$

which can be written

$$\left(EI\alpha^4 + \frac{12Ei}{l^3}\right)w_1 - \frac{12Ei}{l^3}w_2 + 0w_3 = Z_1. \qquad (12.12)$$

Similar equations can be obtained for beams 2 and 3 for equilibrium of bending load on beam, distributing load in planks, and applied load

$$-\frac{12Ei}{l^3} w_1 + \left(\frac{12Ei}{l^3} + EI\alpha^4 + \frac{12Ei}{l^3}\right) w_2 - \frac{12Ei}{l^3} w_3 = Z_2 \quad (12.13)$$

$$0w_1 - \frac{12Ei}{l^3} w_2 + \left(\frac{12Ei}{l^3} + EI\alpha^4\right) w_3 = Z_3. \quad (12.14)$$

Equations (12.12)–(12.14) can be solved to give values of w_1, w_2 and w_3 appropriate to the particular loads Z_1, Z_2 and Z_3. These deflections are then back substituted into equations (12.10) and (12.8) to give the forces in the planks and beams.

The above procedure is followed for every harmonic of the load for which distribution is significant. Then by summing all the harmonics of beam forces, etc., the total distribution of load throughout the deck is obtained.

12.5 FOLDED PLATE ANALYSIS

In the example of Section 12.4, equations (12.12)–(12.14) related the equilibrium of forces on the joint at each beam between the beam itself, the planks spanning to each or one side, and the applied load. The problem was simple because each joint had only one degree of freedom: vertical deflection. In contrast, each longitudinal joint of the folded plate structures in Fig. 12.1(c), (d) and (e) has four degrees of freedom: vertical deflection, rotation about the longitudinal axis, sideways deflection and warping displacement along the line of the joint. All other displacements of a point, such as rotation about the transverse axis, can be thought of as differentials of the four above. Though very much more complicated, the method of analysis is basically the same. The rigorous 'elastic' method of analysis of folded plate structures was derived by Goldberg and Leve [7] and presented in matrix form for computer analysis by DeFries-Skene and Scordelis [8]. Without the shorthand of matrix algebra and the numerical capacity of large computers, accurate solutions would not be practical. For example, with four degrees of freedom per joint, the number of simultaneous equations needed for the analysis of the deck of Fig. 12.1(d) would be at least $14 \times 4 = 56$. In practice it is often convenient to increase the number of plate strips by placing additional longitudinal joints between the slab/web intersections, and the number of simultaneous equations is likely to be closer to 120 for this example.

In addition to having many more systems of forces, the relationships between the forces and displacements in any one plate are much more complex than equations (12.10) or (12.8). Figures 12.11 and 12.12 show the edge forces and displacements on a plate strip separated into those

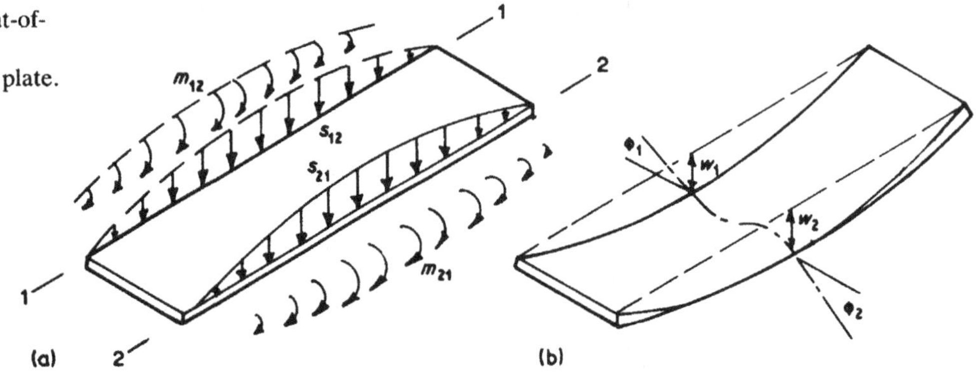

Fig. 12.11 (a) Out-of-plane loads and (b) deflections of plate.

causing out-of-plane flexure and twisting of the plate and those causing in-plane deformation. One of the basic assumptions of folded plate theory is that in-plane and out-of-plane behaviours are independent. A second basic assumption is that the end of every plate is restrained against out-of-plane displacement and rotation (w and ϕ in Fig. 12.11) and against in-plane lateral displacement (v in Fig. 12.12), but is free to warp (u in Fig. 12.12). These support restrictions ensure that the deck is simply supported and that the harmonic analysis has the simplified coefficients described in Section 12.2. Such restrictions on deflection are identical to assuming that at each end of the deck there is a right diaphragm which prevents all displacements within its plane.

The stiffness relationships between the harmonic amplitudes of the forces on one edge of a plate strip and the displacements of the two edges as shown in Figs 12.11 and 12.12 can be written in matrix form:

$$\begin{bmatrix} r_{12} \\ p_{12} \\ s_{12} \\ m_{12} \end{bmatrix} = \begin{bmatrix} g_{11} & k_{11} & 0 & 0 \\ k_{11} & n_{11} & 0 & 0 \\ 0 & 0 & a_{11} & b_{11} \\ 0 & 0 & b_{11} & c_{11} \end{bmatrix} \begin{bmatrix} u_1 \\ v_1 \\ w_1 \\ \phi_1 \end{bmatrix}$$

$$+ \begin{bmatrix} g_{12} & k_{12} & 0 & 0 \\ -k_{12} & n_{12} & 0 & 0 \\ 0 & 0 & a_{12} & b_{12} \\ 0 & 0 & -b_{12} & c_{12} \end{bmatrix} \begin{bmatrix} u_2 \\ v_2 \\ w_2 \\ \phi_2 \end{bmatrix}$$

or

$$\mathbf{r}_1 = \mathbf{k}_{11}\mathbf{u}_1 + \mathbf{k}_{12}\mathbf{u}_2. \tag{12.15}$$

Expressions for the coefficients are given in references [7] and [8] and in the rigorous 'elastic' method every coefficient is complicated. For

Fig. 12.12 (a) In-plane loads and (b) deflections of plate.

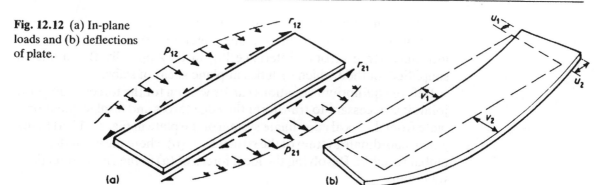

example

$$g_{11} = \frac{Ed}{(1+\nu)^2} \frac{n\pi}{L} \left[- \frac{\sinh\left(\frac{n\pi l}{2L}\right)}{\frac{n\pi l}{2L} \operatorname{csch}\left(\frac{n\pi l}{2L}\right) - \frac{3-\nu}{1+\nu} \cosh\left(\frac{n\pi l}{2L}\right)} + \frac{\cosh\left(\frac{n\pi l}{2L}\right)}{\frac{n\pi l}{2L} \operatorname{sech}\left(\frac{n\pi l}{2L}\right) + \frac{3-\nu}{1+\nu} \sinh\left(\frac{n\pi l}{2L}\right)} \right]$$

(12.16)

where d is the thickness of the plate.

A considerable simplification is introduced to the theory in what is called the 'ordinary' method which is the basis of some computer programs. In this method the plates are treated as transversely spanning strips for out-of-plane flexure, and as simple beams for in-plane

Fig. 12.13 (a) Local and (b) global forces on edge of plate.

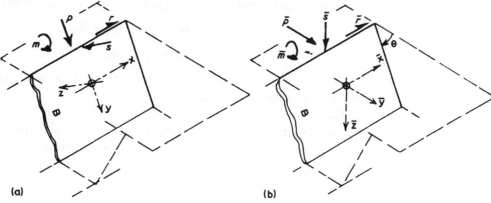

deformation. While for some roof structures little error is introduced, for bridges the neglect of in-plane shear deformation of slabs can introduce large errors. Reference [9], in its Appendix B, outlines a simplified method which includes in-plane shear of slabs.

Before equilibrium equations can be written for the forces acting on a joint, it is necessary to transform the edge forces and displacement on a plate from the local coordinate system of the plate in Fig. 12.13(a) to the global coordinate system of the structure in (b) where the bar indicates a global variable. Resolving the local forces of (a) in the global directions we obtain

$$\begin{bmatrix} \bar{r} \\ \bar{p} \\ \bar{s} \\ \bar{m} \end{bmatrix} = \begin{bmatrix} 1 & 0 & 0 & 0 \\ 0 & \cos\theta & -\sin\theta & 0 \\ 0 & \sin\theta & \cos\theta & 0 \\ 0 & 0 & 0 & 1 \end{bmatrix} \begin{bmatrix} r \\ p \\ s \\ m \end{bmatrix}$$

or

$$\bar{\mathbf{r}} = \mathbf{t}\,\mathbf{r} \qquad (12.17)$$

In addition, compatability of local displacements $u\ v\ w\ \phi$ and global displacements $\bar{u}\ \bar{v}\ \bar{w}\ \bar{\phi}$ requires

$$\begin{bmatrix} u \\ v \\ w \\ \phi \end{bmatrix} = \begin{bmatrix} 1 & 0 & 0 & 0 \\ 0 & \cos\theta & \sin\theta & 0 \\ 0 & -\sin\theta & \cos\theta & 0 \\ 0 & 0 & 0 & 1 \end{bmatrix} \begin{bmatrix} \bar{u} \\ \bar{v} \\ \bar{w} \\ \bar{\phi} \end{bmatrix}$$

or

$$\mathbf{u} = \mathbf{t}^t\,\bar{\mathbf{u}} \qquad (12.18)$$

where t^t is the transpose of t.

From equations (12.16)–(12.18) we can obtain the stiffness relationships between global edge forces and displacements of a plate

$$\bar{\mathbf{r}}_1 = \mathbf{t}\mathbf{k}_{11}\mathbf{t}^t\bar{\mathbf{u}}_1 + \mathbf{t}\mathbf{k}_{12}\mathbf{t}^t\bar{\mathbf{u}}_2$$
$$= \bar{\mathbf{k}}_{11}\bar{\mathbf{u}}_1 + \bar{\mathbf{k}}_{12}\bar{\mathbf{u}}_2. \qquad (12.19)$$

The calculation of the transformation $\mathbf{t}\,\mathbf{k}\,\mathbf{t}^t$ is a relatively simple procedure with modern computers.

Once the stiffness coefficients of all the plates have been calculated in terms of global forces and displacements, all the forces on each joint can be summed to give equilibrium equations such as for joint 2 in Fig. 12.14:

$$\bar{\mathbf{r}}_{21} + \bar{\mathbf{r}}_{23} + \bar{\mathbf{r}}_{24} = \bar{\mathbf{x}}_2 \qquad (12.20)$$

Fig. 12.14 Forces on edges of plates of structure.

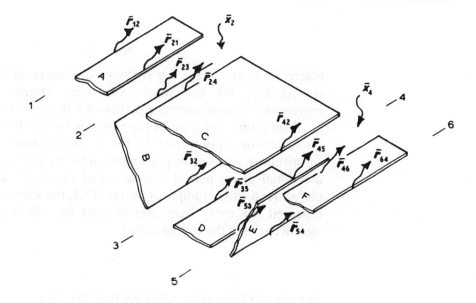

which, using equation (12.19), gives

$$\bar{k}_{A_{21}}\bar{u}_1 + \bar{k}_{A_{22}}\bar{u}_2 + \bar{k}_{B_{22}}\bar{u}_2 + \bar{k}_{B_{23}}\bar{u}_3 + \bar{k}_{C_{22}}\bar{u}_2 + \bar{k}_{C_{24}}\bar{u}_4 = \bar{x}_2$$

or

$$\bar{k}_{A_{21}}\bar{u}_1 + (\bar{k}_{A_{22}} + \bar{k}_{B_{22}} + \bar{k}_{C_{22}})\bar{u}_2 + \bar{k}_{B_{23}}\bar{u}_3 + \bar{k}_{C_{24}}\bar{u}_4 = \bar{x}_2. \quad (12.21)$$

When equation (12.21) is written with the equivalent equations of joints 1, 3, 4, 5 and 6 we obtain

$$\begin{bmatrix} \bar{k}_{A_{11}} & \bar{k}_{A_{12}} & 0 & 0 & 0 & 0 \\ \bar{k}_{A_{21}} & (\bar{k}_{A_{22}} + \bar{k}_{B_{22}} + \bar{k}_{C_{22}}) & \bar{k}_{B_{23}} & \bar{k}_{C_{24}} & 0 & 0 \\ 0 & \bar{k}_{B_{32}} & (\bar{k}_{B_{33}} + \bar{k}_{D_{33}}) & 0 & \bar{k}_{D_{35}} & 0 \\ 0 & \bar{k}_{C_{42}} & 0 & (\bar{k}_{C_{44}} + \bar{k}_{E_{44}} + \bar{k}_{F_{44}}) & \bar{k}_{E_{45}} & \bar{k}_{F_{46}} \\ 0 & 0 & \bar{k}_{D_{53}} & \bar{k}_{E_{54}} & (\bar{k}_{D_{55}} + \bar{k}_{E_{55}}) & 0 \\ 0 & 0 & 0 & \bar{k}_{E_{64}} & 0 & \bar{k}_{F_{66}} \end{bmatrix}$$

$$\times \begin{bmatrix} \bar{u}_1 \\ \bar{u}_2 \\ \bar{u}_3 \\ \bar{u}_4 \\ \bar{u}_5 \\ \bar{u}_6 \end{bmatrix} = \begin{bmatrix} \bar{x}_1 \\ \bar{x}_2 \\ \bar{x}_3 \\ \bar{x}_4 \\ \bar{x}_5 \\ \bar{x}_6 \end{bmatrix}$$

or

$$\bar{K}\bar{u} = \bar{X} \qquad (12.22)$$

where $\bar{K}$, $\bar{U}$ and $\bar{X}$ are the stiffness, displacement and load matrices, respectively, for the whole structure. Matrix equation (12.22) represents 24 simultaneous equations for the sets of four displacements at all six joints. Solution of the equations gives the values of the joint displacements appropriate to the particular load. Subsequent back substitutions of these displacements into the stiffness equations of the individual plates permit calculation of the forces and stresses in the plates. As for the example of Section 12.4, the above procedure must be repeated for every harmonic of load for which the distribution is significant, and the results added.

12.6 CONTINUOUS AND SKEW DECKS

In Section 12.2 it is assumed that the bridge deck is right and simply supported. Then by choosing only harmonics of load which have points of contraflexure coincident with the supports, the bending moment and deflection automatically become zero at the supports. If, on the other hand, the deck has spring supports, internal supports or skew abutments, the analysis becomes very much more cumbersome. If the deck has right abutments but is continuous over an internal support such as J in Fig. 12.15, the deck can still be analysed as if it is simply supported between abutments. First the distributions of moment, deflection, etc. are found independently for a unit reaction $R_J = 1$ and for the live load. Then the distribution due to R_J is scaled up or down so that its deflection

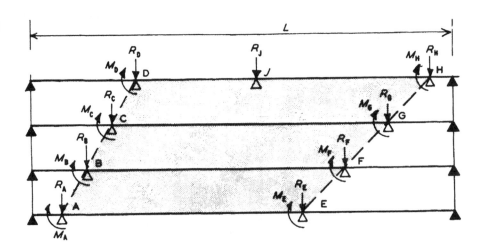

Fig. 12.15 Redundant forces at internal and skew supports.

at J cancels that due to the live load (or for a spring support leaves a residual deflection equal to $R_J \times$ spring stiffness). The distributions are superposed to give the distribution for the continuous deck with no deflection at J. The same principle can be employed for a deck with any support conditions including skew abutments ABCD, EFGH in Fig. 12.15. The skew deck can be thought of as part of a longer structure of arbitrary length L. If the deck has the simple structure of Fig. 12.10, the reactant forces and moments R_A, M_A, etc., are applied to the structure so that the combined distribution of these loads and the live load give no net deflection or moment at each skew support point. More complex structures such as Fig. 12.14 are solved by introducing as many reactant forces at the supports as there are independent deflections or forces which must be eliminated or balanced.

Reference [6] demonstrates a simple technique for analysing a simple skew deck by treating the deck as right and skewing the load.

12.7 ERRORS OF HARMONICS NEAR DISCONTINUITIES

Computer programs employing harmonic analysis can only consider a limited number of harmonics (say ten or one hundred) and ignore the higher harmonics. If the sum of the considered harmonics of load or shear force, etc., is plotted in the region of a discontinuity as in Fig. 12.16, the line violently oscillates close to the discontinuity. Increasing the number of considered harmonics moves the oscillation closer to the discontinuity, but the amplitude is not reduced. At the limit the oscillation still exists but is infinitely narrow. This characteristic of harmonic analysis, called Gibb's phenomenon, can lead to significant errors in the output from folded plate and finite strip computer programs in the region to each side of a discontinuity within two or three wavelengths of the highest harmonic considered. Consequently, unless the computer plots out the oscillating distribution along the deck, results output for discontinuous functions such as shear force over a pier should be considered suspect in the region of the discontinuity. These errors can largely be avoided if the analysis lumps all the undistributed higher harmonics in the loaded member as in Section 12.3.3.

Fig. 12.16 Gibb's phenomenon.

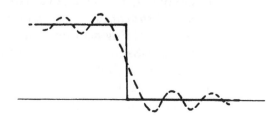

Fig. 12.17 Timber footbridge over River Thames at Temple, England; designed by Groot Lemmer BV, Holland. Photograph courtesy of Sarum Hardwood Structures Ltd.

REFERENCES

1. Kreyszig, E. (1962) *Advanced Engineering Mathematics*, John Wiley, New York.
2. Rowe, R.E. (1962) *Concrete Bridge Design*, CR Books Ltd, London.
3. Hendry, A.W. and Jaeger, L.G. (1958) *The Analysis of Grid Frameworks and Related Structures*, Chatto and Windus, London.
4. Cusens, A.R. and Pama, R.P. (1975) *Bridge Deck Analysis*, John Wiley, London.
5. Bakht, B. and Jaeger, L.G. (1985) *Bridge Analysis Simplified*, McGraw Hill, New York.
6. Jaeger, L.G. and Bakht, B. (1989) *Bridge Analysis by Microcomputer*, McGraw Hill, New York.
7. Goldberg, J.E. and Leve, H.L. (1957) *Theory of prismatic folded plate structures*, International Association for Bridge and Structural Engineering, Zurich, No. 87, pp. 71–2.
8. DeFries-Skene, A. and Scordelis, A.C. (1964) *Direct stiffness solution for folded plates*, Proc. ASCE, **ST 4**, pp. 15–47.
9. Hambly, E.C. (1976) *Bridge Deck Behaviour* (1st edn), Chapman & Hall, London, pp. 288.

13 Finite element method

13.1 INTRODUCTION

The finite element method is a technique for analysing complicated structures by notionally cutting up the continuum of the prototype into a number of small elements which are connected at discrete joints called nodes. For each element, approximate stiffness equations are derived relating the displacements of the nodes to the node forces between elements and, in the same way that slope–deflection equations can be solved for joints in a continuous beam, an electronic computer is used to solve the very large number of simultaneous equations that relate node forces and displacements. Since the basic principle of subdivision of the structure into simple elements can be applied to structures of all forms and complexity, there is no logical limit to the type of structure that can be analysed if the computer program is written in the appropriate form. Consequently, finite elements provide the most versatile method of analysis available at present, and for some structures the only practical method. However, the quantity of computation can be enormous and expensive so that often the cost cannot be justified for run-of-the-mill structures. Furthermore, the numerous different theoretical formulations of element stiffness characteristics all require approximations which in different ways affect the accuracy and applicability of the method. Further research and development is required before the method will have the ease of use and reliability of the simpler methods of bridge deck analysis described in previous chapters.

The technique was pioneered for two-dimensional elastic structures by Turner *et al.* [1] and Clough [2] during the 1950s. Since then a very considerable development has been made by many people. This chapter does little more than demonstrate the basic physical principles. Much more detailed and comprehensive descriptions of the method are given in references [3]–[10].

13.2 TWO-DIMENSIONAL PLANE STRESS ELEMENTS

The finite element method is first demonstrated in relation to the analysis of the plane stress (or 'in-plane' or 'membrane') behaviour of flat plates. This is one of the simplest applications of the method, ignoring continuous beams, and it is relevant to the in-plane actions of the slabs and webs of beam-and-slab and cellular bridge decks. Out-of-plane bending is more complicated and is discussed in Section 13.3.

Figure 13.1(a) shows an elevation of a beam which in (b) is subjected to pure bending. This simple structure is chosen for the example because its behaviour is well known to civil engineers, but the general principles of the following discussion can be applied to a plate of any shape subjected to any system of in-plane forces. For analysis, the structure is considered as in (c) to be made up of a large number of triangular elements connected together only at the corners. The triangular elements are drawn separated except at nodes to emphasize that there is no force interaction along the cuts. When this articulated structure is subjected to pure bending as in (d), the deformation of each element only depends on the movements of the nodes. In the simplest of element models it is assumed that the strains within each element are uniform during distortion with the triangle edges remaining straight, as in Fig. 13.2(a). Thus, while the 45° lines in the prototype of Fig. 13.1(b)

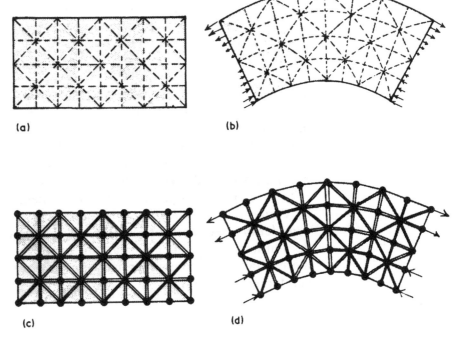

Fig. 13.1 Bending of beam: (a) and (b) prototype; (c) and (d) finite element model.

Fig. 13.2 Node displacements and forces on simple plane stress triangular element: (a) displacements; (b) stresses; and (c) node forces.

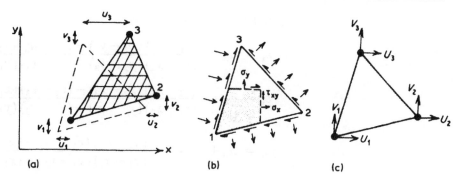

deflect in curves, those in the model of Fig. 13.1(d) deflect as strings of short straight lines. The difference between the model and prototype can be reduced if the model is composed of a larger number of smaller-sized elements. At the theoretical limit with an infinite number of infinitely small elements, the model is effectively a continuum like the prototype.

The stiffness equation of this simple element model can be derived directly from the theory of elasticity plane stress equations. If the triangle of Fig. 13.2(a) is assumed to distort with internal straight lines and edges remaining straight, the displacement field is linear and can be expressed as

$$u = \alpha_1 + \alpha_2 x + \alpha_3 y$$
$$v = \alpha_4 + \alpha_5 x + \alpha_6 y. \tag{13.1}$$

The strains are given by

$$\varepsilon_x = \frac{\partial u}{\partial x} = \alpha_2$$

$$\varepsilon_y = \frac{\partial v}{\partial y} = \alpha_6$$

$$\gamma_{xy} = \frac{\partial u}{\partial y} + \frac{\partial v}{\partial x} = \alpha_3 + \alpha_5. \tag{13.2}$$

If the three nodes have coordinates (x_1, y_1), etc. and displace u_1, v_1, etc. then equations (13.1) can be written for the six displacements. For example, the three equations for u are

$$u_1 = \alpha_1 + \alpha_2 x_1 + \alpha_3 y_1$$
$$u_2 = \alpha_1 + \alpha_2 x_2 + \alpha_3 y_2 \tag{13.3}$$
$$u_3 = \alpha_1 + \alpha_2 x_3 + \alpha_3 y_3$$

which can be solved to give α_1, α_2 and α_3, and with α_4, α_5 and α_6, from the equations for v we obtain

$$\varepsilon_x = \alpha_2 = \frac{(u_1 - u_2)(y_2 - y_3) - (u_2 - u_3)(y_1 - y_2)}{(x_1 - x_2)(y_2 - y_3) - (x_2 - x_3)(y_1 - y_2)}$$

$$\varepsilon_y = \alpha_6 = \frac{(v_1 - v_2)(x_2 - x_3) - (v_2 - v_3)(x_1 - x_2)}{(x_1 - x_2)(y_2 - y_3) - (x_2 - x_3)(y_1 - y_2)}$$

$$\gamma_{xy} = \alpha_3 + \alpha_5 = -\frac{(u_1 - u_2)(x_2 - x_3) - (u_2 - u_3)(x_1 - x_2)}{(x_1 - x_2)(y_2 - y_3) - (x_2 - x_3)(y_1 - y_2)}$$
$$+ \frac{(v_1 - v_2)(y_2 - y_3) - (v_2 - v_3)(y_1 - y_2)}{(x_1 - x_2)(y_2 - y_3) - (x_2 - x_3)(y_1 - y_2)}$$

(13.4)

so relating element strains to node displacements.

The stresses in the element can be found from the elastic theory equations for plane stress

$$\sigma_x = \frac{E}{(1-v^2)}\varepsilon_x + \frac{vE}{(1-v^2)}\varepsilon_y$$

$$\sigma_y = \frac{vE}{(1-v^2)}\varepsilon_x + \frac{E}{(1-v^2)}\varepsilon_y \qquad (13.5)$$

$$\tau_{xy} = \frac{E}{2(1+v)}\gamma_{xy}.$$

If the element were part of a continuum, stresses of this magnitude would cross the boundaries as shown in Fig. 13.2(b). However, these stresses are represented by the node forces of (c) with the stresses on each edge statically distributed to the neighbouring nodes. Since the stresses forming U_1 and V_1 are in equilibrium with the stresses

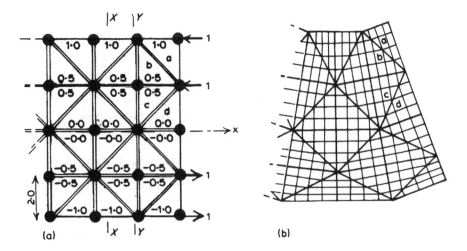

Fig. 13.3 Pure bending of triangular mesh: (a) distribution of σ_x and (b) displacement field.

$(\sigma_x, \sigma_y, \tau_{xy})$ on the internal cuts shown through the midpoints of the sides we can resolve

$$U_1 = \sigma_x \frac{(y_2 - y_3)}{2} - \tau_{xy} \frac{(x_2 - x_3)}{2}$$

$$V_1 = -\sigma_y \frac{(x_2 - x_3)}{2} + \tau_{xy} \frac{(y_2 - y_3)}{2}.$$

(13.6)

Substituting for σ_x, etc. from equation (13.5),

$$U_1 = \frac{E}{(1-v^2)} \frac{(y_2 - y_3)}{2} \varepsilon_x + \frac{vE}{(1-v^2)} \frac{(y_2 - y_3)}{2} \varepsilon_y$$
$$- \frac{E}{2(1+v)} \frac{(x_2 - x_3)}{2} \gamma_{xy}$$

$$V_1 = -\frac{vE}{(1-v^2)} \frac{(x_2 - x_3)}{2} \varepsilon_x - \frac{E}{(1-v^2)} \frac{(x_2 - x_3)}{2} \varepsilon_y$$
$$+ \frac{E}{2(1+v)} \frac{(y_2 - y_3)}{2} \gamma_{xy}.$$

(13.7)

These equations can be combined with equation (13.4) to give stiffness equations for node 1 as a point of the element

$$U_1 = k_1 u_1 \qquad (13.8)$$

relating forces on node 1 to displacements of nodes 1, 2 and 3. Then consideration of the equilibrium of node 1 under the sum of forces from all elements adjoining node 1, together with applied loads X_1, provides stiffness equations for node 1, as a point in the structure

$$X_1 = K_1 \text{[displacements of node 1 and neighbours]}. \qquad (13.9)$$

Such equations can be derived for every node in the model structure, forming in all $2N$ simultaneous equations for N nodes. These equations are solved by the computer to give the displacements (u, v) at every node. From these the strains and stresses in every element can be calculated using equations (13.4) and (13.5).

An example of the stress distribution computed from a coarse mesh model subjected to pure bending is shown in Fig. 13.3(a). The associated distortion of the elements is shown in (b). Since the stresses are only induced here in the x-direction, the stress in each element is simply proportional to the shortening of the edge parallel to the x-axis. Thus it can be seen in (b) that the x-compressions of elements b and c are the same and equal to half that of a, while d is unstrained. Figure 13.4(a) shows how the stresses in elements a, b, c and d from node forces. The

Fig. 13.4 (a) Stresses and node forces on elements and (b) stress distributions on sections.

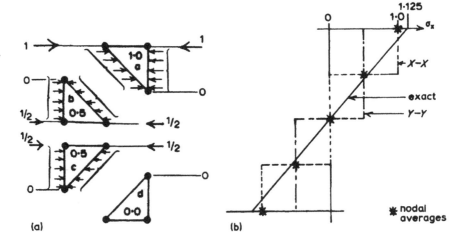

forces are zero where the stresses on the edges on the two sides of the node cancel out. Figure 13.4(b) shows the stress distributions that could be plotted for sections X–X and Y–Y in Fig. 13.3(a). These can be compared with the exact stress distribution. It is evident that the stresses computed for individual elements are here misleading. However the average stresses round each node provide an estimate within 13% of the exact figure.

For the more complex elements, it is not usually possible to relate node forces directly to element stresses and the assumed displacement field. The element stiffnesses are frequently derived from a consideration of the potential energy stored by the assumed displacement field. For each node force, the external work done during a virtual displacement is equated to the minimum increase in potential energy that can be stored by the displacement field. While this mathematical concept enables more complicated problems to be analysed with computer economy, it does mean that element stresses output by a program are not directly related by equilibrium to the applied loads. If the element displacement function is not appropriate to the problem, the output element stresses can be as low as 50% or less of those necessary for equilibrium with the applied loads. Calculating the nodal averages does not necessarily make much difference. Although continual improvements in programs are reducing the possibility of such errors, it is advisable to make a hand check of the equilibrium of output element stresses and the applied loads wherever possible. Because significant discrepancies have been found in the past, many design engineers still place more confidence in a space frame analysis in which output member forces are automatically in equilibrium with applied loads.

Fig. 13.5 More accurate stresses obtained with finer mesh.

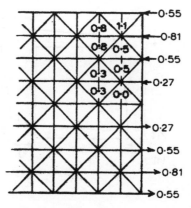

A finer mesh is shown in Fig. 13.5 and it is evident that the nodal average stresses are now within 3% of the exact figures in Fig. 13.4. In general, the finer the finite element mesh is, the more accurate are the results. A very coarse mesh, as shown in Fig. 13.6, can be extremely inaccurate. Furthermore, the curvature of the beam, proportional to the difference between the top and bottom element stresses, is very much less than that for Fig. 13.5. This implies that the coarse mesh is very much stiffer than the fine mesh. It might appear foolish even to consider a mesh as coarse as Fig. 13.6(a). However, the example is a warning of the very large errors that might be introduced if this beam were a web of a box-girder such as in (b) where at a first glance the mesh might not appear coarse. The problem highlights the importance of choosing the most appropriate element arrangement and of checking the solution when possible against a different type of analysis.

Fig. 13.6 (a) Inaccurate stresses from very coarse mesh and (b) coarse mesh of box girder.

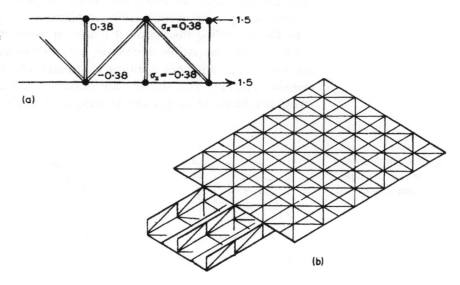

Fig. 13.7 Node displacements and forces on quadrilateral element.

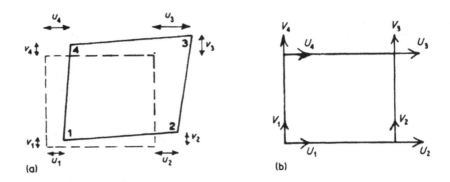

The finite elements of a model do not have to be triangular; numerous other shapes including quadrilaterals, rectangles, parallelograms and other polygons are used. Figure 13.7 shows the node displacements and forces of a rectangular element. The displacements can be assumed to have functions

$$u = \alpha_1 + \alpha_2 x + \alpha_3 y + \alpha_4 xy$$
$$v = \alpha_5 + \alpha_6 x + \alpha_7 y + \alpha_8 xy.$$

(13.10)

The distortion is then similar to Fig. 13.8(a) with linear displacements (i.e. constant strain) along each edge. It can be noted that equations (13.10) have eight unknowns which are found by solving the equations for the eight displacements $u_1, v_1, u_2, \ldots, v_4$. An alternative approach is to assume that the rectangle has the stiffness characteristics of the pairs of triangles of either (b) or (c) or of the average of (b) and (c). Other rectangular elements have also been derived for various purposes using different displacement functions. Some, in particular [8], have been evolved to represent webs bending in-plane.

In the preceding discussion it has been assumed that the model is made up of compatible elements, i.e. displacement functions are assumed so that adjacent points on elements on each side of a cut always remain adjacent. The stresses are discontinuous at the cuts, but the stress resultants at nodes are in equilibrium. By making the model

Fig. 13.8 Displacement fields in various simple quadrilateral elements.

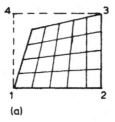
(a)

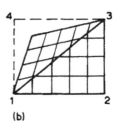
(b)

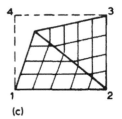
(c)

Fig. 13.9 Finite element model for plate bending of slab deck.

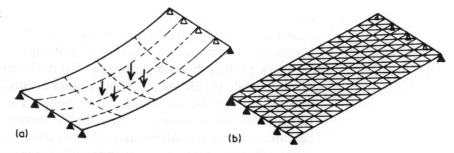

distort in a specified way, it has less freedom and is stiffer than the prototype and computed stresses are lower bounds, always being less than exact solutions. This was demonstrated in Fig. 13.4. An alternative approach is to assume stress functions for the elements so that stresses are continuous across the edges of elements. Displacements are then discontinuous with relative displacements between edges of elements except at the nodes. The resulting model is then more flexible than the prototype and computed stresses are upper bounds, being greater than the exact ones. This flexibility method is not used very often because of its greater theoretical complexity.

13.3 PLATE BENDING ELEMENTS

The finite element technique is often used for analysis of plate bending behaviour of slab bridges. The basic concept and process is similar to

Fig. 13.10 (a) triangular plate element; (b) edge stress resultants; (c) node displacements; and (d) node forces.

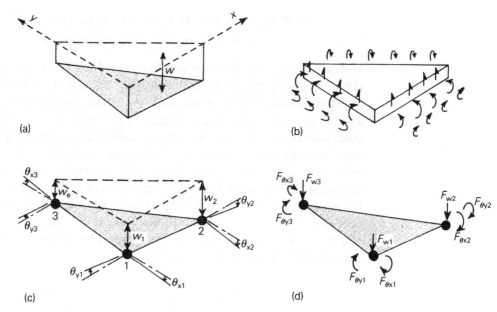

that for plane stress. However the deflection and force variables are different, and the theoretical derivation of the element stiffness equations presents greater theoretical problems.

Figure 13.9(a) shows a prototype slab deck bending under vertical loading, while (b) shows a possible system of subdivision of the model into a large number of triangular elements. The piece of prototype represented by an element experiences vertical deflection w (and curvatures which are differentials of w) as shown in Fig. 13.10(a) and is subjected to moments, torsions and vertical shear forces along its edges, as in (b). These edge forces are represented in the stiffness equations by the node forces shown in (d):

$$F_w = \text{a vertical shear force}$$
$$F_{\theta x} = \text{a moment about axis } Ox$$
$$F_{\theta y} = \text{a moment about axis } Oy$$

The node deflections in (c) appropriate to the forces in (d) are

$$w = \text{vertical deflection}$$
$$\theta_x = \text{rotation of node about } Ox$$
$$\theta_y = \text{rotation of node about } Oy.$$

To determine the stiffness of such an element, a displacement field must be assumed as was done for plane stress elements in equation (13.1). A first choice might be the complete third-order polynomial:

$$w = \alpha_1 + \alpha_2 x + \alpha_3 y + \alpha_4 x^2 + \alpha_5 xy + \alpha_6 y^2$$
$$+ \alpha_7 x^3 + \alpha_8 x^2 y + \alpha_9 xy^2 + \alpha_{10} y^3. \qquad (13.11)$$

Unfortunately, this has ten coefficients which cannot be found from the nine displacement variables w_1, θ_{x_1}, θ_{y_1}, etc. for the element. An approximation must be made and one coefficient removed somewhat arbitrarily, or two made equal ($\alpha_8 = \alpha_9$). (This can lead to computational problems, and often a different coordinate system called area coordinates is used for triangular elements, as described in reference [4].) Along any edge, for example with $y = $ constant, equation (13.11) gives w as a cubic function of x; this will be the same on both sides of an interface since the cubic with four coefficients is uniquely defined by four node displacements. However the normal slope $\partial w/\partial y$ is a quadratic of x with three coefficients, and since there are only two defining node variables

$$\theta_{x_1} = \left(\frac{\partial w}{\partial y}\right)_1 \quad \text{and} \quad \theta_{x_2} = \left(\frac{\partial w}{\partial y}\right)_2$$

Fig. 13.11 Discontinuity of slope at element edges between nodes.

on the edge, the normal slope is not uniquely defined and can be different for the elements on two sides of an interface. Consequently, as is shown in Fig. 13.11, except at nodes, there is a kink at the interface between elements even though the vertical deflections are continuous. Thus, because the complete polynominal is too general for the number of defining degrees of freedom, the elements are not truly compatible, and are said to be 'non-conforming'.

A simple rectangular element such as Fig. 13.12(a) is also non-conforming, since the lowest order polynomial with sufficient coefficients to describe the deflections is fourth order with sixteen coefficients while the element has only twelve defining degrees of freedom. Numerous other bending elements have been proposed which are also non-conforming. However there are some, such as in Fig. 13.12(b), which are made 'conforming'. In this example, the number of defining degrees of freedom of the element is increased by adding additional nodes at the midpoints of each edge. At these nodes the normal slope ($\partial n/\partial s$) is the displacement variable, with bending moment the node force. Now a complete fourth-order polynomial with sixteen coefficients can be solved using the sixteen degrees of freedom. The normal slope which varies as a quadratic along any edge is uniquely defined by three values at the two end nodes and one middle node.

The stiffness equations for the elements are usually derived, as mentioned in Section 13.2, by a consideration of virtual work. The node is given a virtual displacement for each of the degrees of freedom, and the corresponding node forces are found by equating the externally applied work to the minimized increase in potential energy stored by the assumed displacement function.

Fig. 13.12 Quadrilateral plate bending elements: (a) non-conforming and (b) conforming.

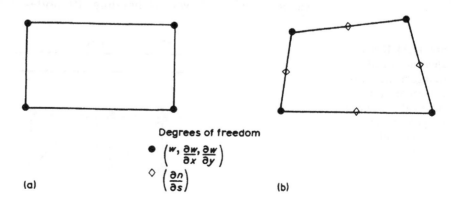

Fig. 13.13 Slab deck divided into triangular elements.

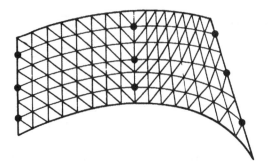

Solutions obtained with compatible conforming elements are always lower bounds since the model structure is stiffer when the ways it can deform are completely specified. In contrast, solutions obtained with non-conforming elements are neither lower bounds nor upper bounds. They do not satisfy either of the following assumptions: compatibility throughout the structure for a lower-bound minimum potential energy solution; or equilibrium throughout the structure for an upper-bound minimum complementary energy solution. A non-conforming element will give higher stresses than the lower bound obtained from a conforming element of the same shape because it has less stiffness due to its greater freedom in taking up a deflected shape. Sometimes, ironically, the simple non-conforming triangular or rectangular elements can give better solutions than much more sophisticated conforming elements. This is done at a very great saving in cost since the conforming elements are significantly more cumbersome for computer manipulation. In consequence, non-conforming triangular or quadrilateral elements are frequently used for slab bridges. The triangle is popular because it can be made to fit decks of complex plan shape as in Fig. 13.13.

Before leaving bending elements, it is worth mentioning the simple beam element which, as shown in Fig. 13.14, can be used to represent the stiffening due to a beam in the plane of the slab. Since the beam carries torsion as well as bending, its nodes have three degrees of

Fig. 13.14 Beam element located between triangular elements: (a) section of slab stiffened by beam and (b) plan of finite elements with beam element.

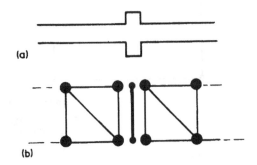

freedom, just like the plate nodes, and the node forces are vertical shear, bending moment and torsion. It is the basic element of a grillage model, which itself can be thought of as a simple finite element model. It should be noted that the in-plane beam of Fig. 13.14 is no better than a grillage beam at representing the slab–membrane action due to a beam located below the slab of a beam-and-slab deck. Consequently, it is unlikely that a two-dimensional plate bending finite element analysis will produce more reliable results than a grillage for such a deck. Only a three-dimensional analytical model could produce more accurate predictions. On the other hand, a two-dimensional finite element analysis with fine mesh should produce more accurate results than a grillage for a plane slab deck when Poisson's ratio is large so that there is interaction in the plate bending equations (3.5) or (3.9).

13.4 THREE-DIMENSIONAL PLATE STRUCTURES AND SHELL ELEMENTS

A detailed investigation of a beam-and-slab or cellular bridge deck requires a three-dimensional analysis. Even though it is generally possible to approximate the behaviour of slabs and webs to thin plates, these must be arranged in a three-dimensional assemblage as in Fig. 13.15.

At every intersection of plates lying in different planes there is an interaction between the in-plane forces of one plate and the out-of-plane forces of the other, and vice versa. For this reason it is essential to use finite elements which can distort under plane stress as well as plate bending. Since it is assumed that for flat plates, in-plane and out-of-plane forces do not interact within the plate, the elements are in effect

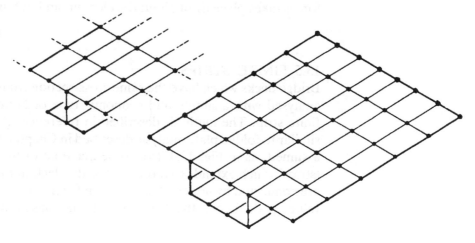

Fig. 13.15 Three-dimensional structures composed of plate elements.

Fig. 13.16 Arched structures of (a) plate elements and (b) shell elements.

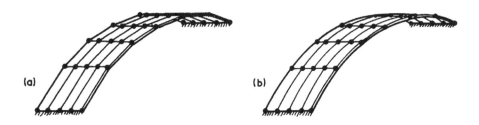

the same as a plane stress element (described in Section 13.2) in parallel with a plate bending element of Section 13.1.

It should be noted that even if the elements are conforming individually for plane stress and bending, the displacements are not usually compatible along the web/slab intersections except at nodes. For the reasons discussed in Section 13.2, the simple triangular element is not suitable for such an analysis unless an extremely fine mesh is used. However special elements are available which can represent the in-plane bending of the webs.

There is no logical limit to the cellular complexity, structural shape or support system of a bridge that can be analysed with a three-dimensional plate model. However every node must have six degrees of freedom, three deflections and three rotations, and solution of the very large number of stiffness equations generated for even relatively simple structures is expensive. Consequently, the method is usually only used to study the distribution of stress in one span or an intricate part of a structure. These results are then used to interpret the distributions of stress resultants such as the overall moment that are output by the simpler models of continuous beam, grillage or space frame.

Shell structures such as arches can usually be analysed with plate elements as in Fig. 13.16(a). However, shell elements as in (b) are available. In such elements the interaction of in-plane and out-of-plane forces takes place throughout the element and not just at the nodes.

13.5 FINITE STRIPS

Bridge decks which have the same cross-section from end to end can be analysed with a simple and economic type of finite element called a finite strip. The method, described in references [9] and [10], is very similar to folded plate analysis described in Chapter 12. The structure is assumed, as in Fig. 13.17(a), to be made up of finite elements called 'strips' which extend from one end of the deck to the other. The strips are connected by nodes which also run from one end to the other. Like folded plate theory, the displacement functions for in-plane and out-of-

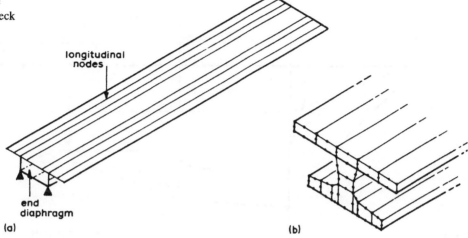

Fig. 13.17 (a) Finite strip model of box deck and (b) finite prism model.

plane deformation of the strips are of the form

$$w, \theta, u \text{ or } v = \Sigma f(y) \sin\left(\frac{n\pi x}{L}\right) \quad (13.12)$$

where x is the direction along the structure and y is the direction across the strip. As explained in Chapter 12, harmonic analysis is greatly simplified if the deck is assumed to have right end supports with diaphragms to prevent displacement of the ends of the plates in the plane of the diaphragm. The analytical procedure is also the same in that stiffness equations are obtained and solved for each harmonic component of the load in turn, and the results summed to give the total stress distribution. Furthermore, similar errors due to Gibb's phenomenon can be encountered near discontinuities (see Section 12.7). In the finite strip method, the transverse functions $f(y)$ are assumed to be simple polynomials so that in effect the method is an approximation to the rigorous 'elastic' folded plate method in which these functions have a complicated hyperbolic form similar to equation (12.16). One result of this approximation is that calculated stresses are discontinuous transversely at strip interfaces. Figure 13.18 shows the in-plane shear stress distribution output for a two-cell spine which is part of a two-spine concrete deck. The computer output indicates discontinuities of shear flow at the strip interfaces within slabs. These discontinuities, which are physically impossible, must be smoothed out to a 'sensible fit'. Also shown for comparison are the shear flows calculated from a grillage following the procedures of Chapter 5.

Harmonic analysis can also be used with two-dimensional displacement functions to analyse prismatic solid structures such as Fig.

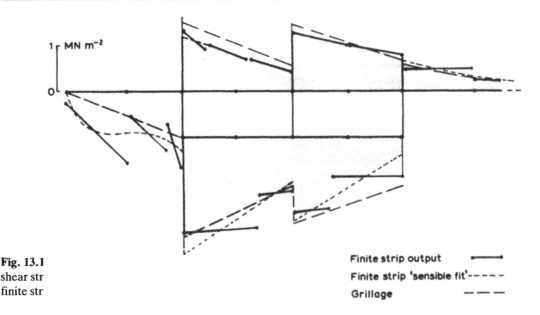

Fig. 13.1
shear str
finite str

13.17(b). Furthermore, this technique and finite strip analysis have been developed for curved circular structures with the harmonic function used for variations along circular arcs.

13.6 THREE-DIMENSIONAL ELEMENTS

Three-dimensional solid elements are seldom used in the analysis of bridge decks because generally these structures are composed of the thinnest plates possible to minimize weight. Solid elements are used more often for the analysis of nuclear reactors and complex soil structures. The simplest elements consist of tetrahedra or hexahedra with nodes at the corners, as shown in Fig. 13.19(a) and (b). If the mesh is fine, the nodes need only have three degrees of freedom for displacement in the three dimensions. More sophisticated elements have additional nodes in addition to those at the corners, with more degrees of freedom at each node.

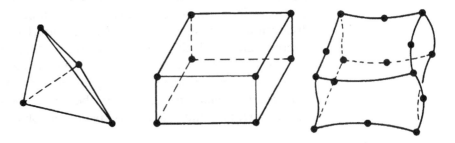

Fig. 13.19 Three-dimensional sold elements.

13.7 CONCLUSION

The finite element method is the most powerful and versatile analytical method available at present because with a sufficiently large computer, the elastic behaviour of almost any structure can be analysed accurately. For this reason it is often requested by clients, or proposed to a client, to show that the most accurate analysis possible has been performed. Unfortunately, the method is cumbersome to use and is usually expensive. In addition, the choice of element type can be extremely critical and, if incorrect, the results can be far more inaccurate than those predicted by simpler models such as grillage or space frame. However, perhaps the greatest drawback at present is that while the technique is developing so rapidly, the job of carrying out finite element computations is a full time occupation which cannot be carried out at the same time by the senior engineer responsible for the design. He is unlikely to have time to understand or verify the appropriateness of the element stiffnesses or to check the large quantity of computer data. This makes it difficult for him to place his confidence in the results, especially

Fig. 13.20 Twin-cell concrete box-girders of Redbridge Flyover, Southampton, England; designed by Gifford & Partners. See Fig. 13.18. Photograph E.C. Hambly.

if the structure is too complicated for him to use simple physical reasoning to check orders of magnitude.

For these reasons, when commissioning a finite element analysis, it is advisable to check that the computer organization and employees involved have had plenty of experience of the program, and that the program has been well tested for similar structural problems. Furthermore, if the structure is too complicated to apply physical reasoning, it is worth commissioning an inexpensive simple frame analysis as an independent check.

REFERENCES

1. Turner, M.J., Clough, R.W., Martin, H.C. and Topp, L.J. (1956) Stiffness and deflection analysis of complex structures, *J. Aero. Sci.* **23**, 805–23.
2. Clough, R.W. (1960) 'The finite element in plane stress analysis', Proc. 2nd A.S.C.E. Conf. on Electronic Computation, Pittsburg, Pa., Sept.
3. Irons, B. and Shrive, N. (1983) *Finite Element Primer*, Ellis Horwood, Chichester, and John Wiley, New York.
4. Zienkiewicz, O.C. and Taylor, R.L. (1989) *The Finite Element Method*, McGraw-Hill, London, 4th edn.
5. Irons, B. and Ahmad, S. (1980) *Techniques of Finite Elements*, Ellis Horwood, Chichester, and John Wiley, New York.
6. Burnett, D.S. (1987) *Finite Element Analysis*, Addison-Wesley, Reading, Mass.
7. Desai, C.S. and Abel, J.F. (1972) *Introduction to the Finite Element Method*, Van Nostrand Reinhold, New York.
8. Rockey, K.C., Evans, H.R., Griffiths, D.W. and Nethercot, D.A. (1983) *The Finite Element Method*, Collins, London, 2nd edn.
9. Cheung, Y.K. (1968) The finite strip method in the analysis of elastic plates with two opposite simply supported ends, *Proc. Inst. Civ. Eng.*, **40**, 1–7.
10. Loo, Y.C. and Cusens, A.R. (1978) *The Finite Strip Method in Bridge Engineering*, Viewpoint Publication (now E. & F.N. Spon), London.

14 Stiffnesses of supports and foundations

14.1 INTRODUCTION

Displacements of supports under load can make a large difference to the distribution of forces within a bridge deck. For this reason it is important to devote as much care to the assessment of the stiffnesses of the supports and foundations as to the stiffnesses of the deck structure. The influence of bearing stiffnesses and foundation settlement is illustrated in Fig. 14.1 for a skew slab deck with two spans of 22.5 m continuous over the pier. Figure 14.1(b) illustrates the bending moment diagram calculated by a grillage for dead load, when the structure is supported on rigid supports after the falsework has been removed from under the two spans. Figure 14.1(c) illustrates the moments due to dead load when it coexists with a differential settlement of 25 mm along the length of the pier (14 m). Figure 14.1(d) shows the moments that are calculated when the supports of the grillage are given the stiffnesses of rubber bearings. It is evident that the maximum moment over a bearing in Fig. 14.1(b) is

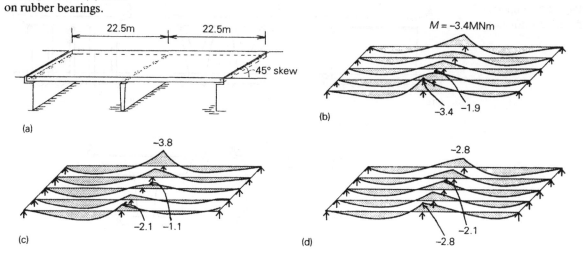

Fig. 14.1 Moments due to distributed load on two-span skew slab with various support conditions: (a) arrangement; (b) deck on rigid supports; (c) with 25 mm differential settlement across pier; and (d) deck on rubber bearings.

3.4 MNm, while it is 3.8 MNm in (c) and 2.8 MNm in (d). The reactions on the bearings differ in a similar manner.

The distribution of forces throughout a structure, and between supports, depends on the relative stiffnesses of all the components of the structure–foundation system. Different relative stiffnesses are often needed for different parts of the calculations. The change in moments from Fig. 14.1(b) and Fig. 14.1(c) depends on the movement of the whole line of supports relative to the deck flexing from end to end. The differences between Fig. 14.1(b) and Fig. 14.1(d) depend on the stiffness of individual bearings as compared to the deck spanning transversely between bearings.

Continuous structures can be very effective at distributing forces, but they are also very sensitive to the effects of compressibility of supports and foundations. A box-girder which is very stiff against torsion and distortion can be very sensitive to differential settlement and compression of bearings. If it has two bearings close together at a support it may only require a small differential settlement, or small twist of the structure, for the whole reaction to pass through one bearing.

14.2 SUBSTRUCTURES AND BEARINGS

Substructures can be very stiff under vertical load, particularly if they transmit loads by direct compression forces and not by bending. Figure 14.2(a), (b) and (c) illustrate piers which can transmit deck loads to the foundations by direct compression. In contrast the supports of Fig. 14.2(d) and (e), which involve a cross-head beam or cantilever, may have significant flexibility. If a bridge deck on such supports is analysed with a grillage then it may be necessary to model the cross-head as well

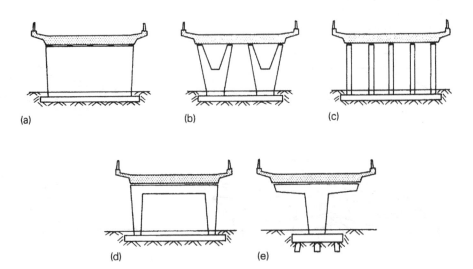

Fig. 14.2 Supports of differing stiffnesses.

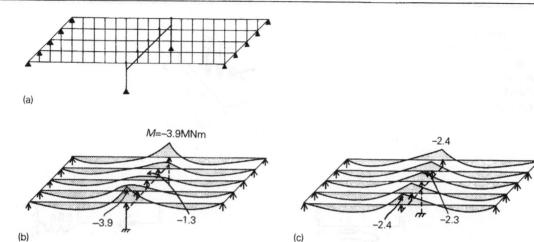

Fig. 14.3 Moments in deck of Fig. 14.1 with different types of pier: (a) space frame model; (b) portal frame pier; and (c) cantilever cross-head pier.

as the deck. It may be easier to model the structure with a space frame which reproduces the shape of the pier. It is then relatively simple to model compressible bearings with vertical members between deck and cross-head, and to model the foundation stiffnesses. Figure 14.3(a) illustrates a space frame model for the deck of Fig. 14.1 supported on the pier of Fig. 14.2(d).

Figure 14.3(b) and (c) illustrate the influence of different types of support on the bending moments in the slab deck of Fig. 14.1 under dead load. Figure 14.3(b) shows the moments when the deck is supported on a cross-head beam, as in Fig. 14.2(d), with stiff bearings. Figure 14.3(c) shows the moments when the deck is supported on the cantilever cross-head of Fig. 14.2(e) with rubber bearings. It is evident that the maximum moment over the pier in Fig. 14.3(b) (3.9 MNm) is 60% greater than the value in Fig. 14.3(c) (2.4 MNm). The deck and loading is the same in the two examples, and the large difference in hogging moments is due to the different support conditions. However the load distribution in a beam-and-slab deck is not so sensitive to differences in support stiffnesses because the transverse bending stiffness of the slab is much lower.

14.3 FOUNDATION STIFFNESSES

Approximate estimates for the stiffnesses of shallow footing foundations, shown in Fig. 14.4, can be obtained with the equations in references [1] and [2] based on the theory for an elastic half space. The various equations in references [1] and [2] can be simplified to the following approximate forms when the footings slide or tilt across the shorter direction.

Fig. 14.4 Stiffnesses of footings: (a) perspective; (b) vertical; (c) horizontal; (d) rotational; and (e) combined.

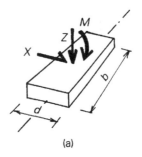

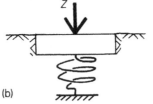

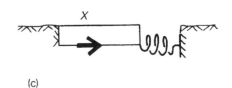

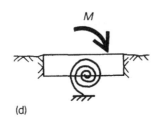

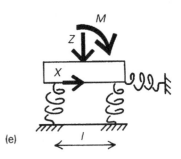

$$\text{shear modulus} \quad G = \frac{E}{2(1+\nu)} \quad (14.1)$$

$$\text{vertical stiffness} \quad K_z = \frac{2.5GA^{0.5}}{(1-\nu)} \quad (14.2)$$

$$\text{horizontal stiffness} \quad K_x = 2G(1+\nu)A^{0.5} \quad (14.3)$$

$$\text{rocking stiffness} \quad K_m = \frac{2.5GZ}{(1-\nu)} \quad (14.4)$$

where

G = shear modulus of soil
E = Young's modulus of soil
ν = Poisson's ratio of soil
A = foundation area $= bd$
Z = foundation section modulus $= bd^2/6$.

If ν is of the order of 0.3 to 0.5 these equations can be approximated by

$$K_z = 1.5EA^{0.5} \qquad (14.5)$$

$$K_x = EA^{0.5} \qquad (14.6)$$

$$K_m = 1.5EZ. \qquad (14.7)$$

It is sometimes convenient to represent the vertical and rotational stiffnesses by two parallel springs spaced l apart as in Fig. 14.4(e). The stiffness and spacing of the vertical springs are given by

$$K = 0.5K_z \quad l = 2(K_m/K_z)^{0.5} = 0.82b^{0.25}\,d^{0.75}. \qquad (14.8)$$

Real ground conditions often differ markedly from the simplification of an elastic half space. In addition the above equations give no indication of the interaction of the stiffnesses under complex loading; a matrix of stiffnesses is really required. Consequently the above equations should be used with care. None the less they provide a quick means of determining the order of magnitude of foundation stiffnesses.

The stiffness of the ground under a foundation depends on the paths of reacting forces, as illustrated in Fig. 14.5. The vertical loads on the foundations in Fig. 14.5(a) react with gravitational forces at great depth. The vertical movement of one foundation is slightly increased by the loading on the other foundation. In contrast under arching action the horizontal force and moment on one foundation is balanced by an equal and opposite force and moment from the other foundation. If the foundations are relatively close together the horizontal stiffnesses and

Fig. 14.5 Load paths in the ground.

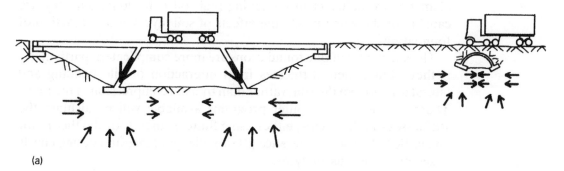

(a)

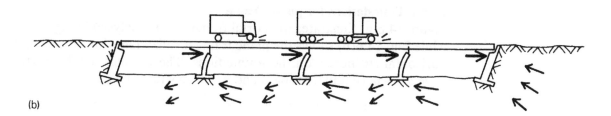

(b)

moment stiffnesses may be increased by the shortness of the load path through the ground between them as compared to the half space at distance. However horizontal loads from braking in Fig. 14.5(b) apply unbalanced reactions to the foundations; so that the horizontal stiffness of any foundation is reduced by concurrent loading on neighbouring foundations. It is often impracticable to take account of these subtleties when selecting foundation stiffnesses, but it is sensible to consider possible load paths of balancing forces. The horizontal stiffnesses of foundations can sometimes be increased in the design by including a compacted layer of fill, or by ground beams, or ties, or by piles.

Detailed investigations of the stiffnesses of foundations for bridges over complicated ground conditions can be carried out with three-dimensional finite element analyses. These analyses have proved to be useful, not only for complicated ground conditions, but also as occasional calibrations for the simple equations such as (14.2) to (14.7). It is preferable to represent the ground as a three-dimensional solid rather than as a two-dimensional vertical plane, because isolated foundations gain much of their stiffness from the lateral spread of forces.

Finite element programs are available which can represent the soils with very sophisticated non-linear stress–strain behaviour with coexisting porewater pressures. While sophisticated programs have greater realism than simpler programs and are very useful for extending our understanding of the phenomena, they require soil data of a quality that is seldom available, and they are unwieldy for most design situations. On the other hand the simplest of three-dimensional finite elements are deficient in modelling real soil behaviour, but they are capable of showing quickly the effects of soil layers and interaction of foundations.

The stiffnesses of pile foundations are more complicated, particularly if they obtain their stiffnesses from interaction of pile bending and lateral forces from the soil, rather than by axial compression of the piles. There are several computer programs available which calculate the stiffnesses of pile groups, either using finite elements, or the theory for an elastic half space. Reference [3] provides guidance and design charts based on numerous analyses.

14.3.1 Foundations for portal frame

Figure 14.6(a) illustrates the portal frame that is analysed in Fig. 2.29, while Fig. 14.6(b) shows the plane frame idealization with vertical and horizontal stiffnesses on the foundations. The footings for the portal have $d = 4$ m (parallel to span) and $b = 16$ m wide. Hence

$$A = 16 \times 4 = 64 \text{ m}^2 \quad Z = 16 \times 4^2/6 = 43 \text{ m}^3$$

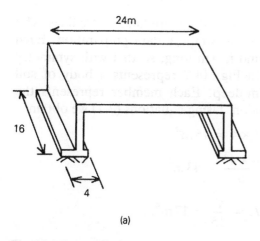
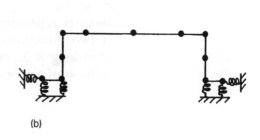

Fig. 14.6 Portal bridge: (a) arrangement and (b) plane frame model.

and equations (14.5), (14.6) and (14.7) give

$$K_z = 1.5\, E\, 64^{0.5} = 12E$$
$$K_x = E\, 64^{0.5} = 8E$$
$$K_m = 1.5\, E\, 43 = 64E.$$

If $E = 100$ MPa we find $K_z = 1200$ MN/m, $K_x = 800$ MN/m and $K_m = 6400$ MNm. It should be noted that because of the square root in equations (14.5) and (14.6) these stiffnesses have to be calculated for the full foundation before being reduced to stiffness per unit width.

The analysis for Fig. 2.29 is carried out with a plane frame representing a 1 m width of the portal structure. Accordingly the above stiffnesses are reduced to stiffnesses per unit width of

$$K_z = 1200/16 = 75 \text{ MN/m/m}$$
$$K_x = 800/16 = 50 \text{ MN/m/m}$$
$$K_m = 6400/16 = 400 \text{ MN/m/rad/m}.$$

Stiffnesses K_z and K_m are represented by two parallel springs of

$$K = 0.5 K_z = 38 \text{ MN/m/m}$$

at spacing

$$l = 2(400/75)^{0.5} = 4.6 \text{ m}.$$

The influence of more complicated ground conditions can be investigated with a finite element analysis. Any degree of sophistication can be included. However, to demonstrate how basic a three-dimensional model can be, Fig. 14.7 illustrates a space frame idealization using the cruciform frame method described in Chapter 7. The model represents one quarter of the portal structure, with symmetry assumed about transverse and longitudinal centre lines. The model has a regular cubic frame to simplify data preparation. Each member representing the ground is 4 m long, so that with symmetry in two directions the model in Fig. 14.7 represents a body of soil 64 m long × 32 m wide × 20 m deep. Each member represents the stiffness of soil of 4 m × 4 m section. Using equations (7.1) we obtain:

$$A_x = 4 \times 4 = 16 \, \text{m}^2$$

$$C_x = \frac{4 \times 4^3}{6} = 43 \, \text{m}^4$$

$$I_y = I_z = \frac{4^4}{15} = 17 \, \text{m}^4.$$

Fig. 14.7 Cruciform space frame idealization of quarter of portal bridge and large volume of ground below.

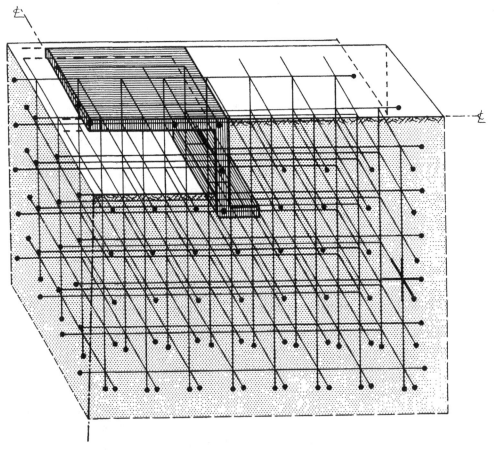

The portal frame has a half width of 8 m and is 1.2 m thick with $E = 30\,000$ MPa. The ground substrata are assumed to have $E = 100$ MPa. The embankment is arbitrarily given $E = 50$ MPa, i.e. half that of the ground, but it is found to have little influence here on arching action. Symmetry and antisymmetry can be considered separately for the analysis of balanced and unbalanced loads.

Figure 14.8 illustrates the deflected shape and the line of thrust derived from the bending moments when the bridge supports a uniformly distributed load. It is found from the reactions and displacements due to the distributed load that the equivalent average ground stiffnesses are here approximately: $K_z = 80$ MN/m/m, $K_x = 90$ MN/m/m and $K_m = 600$ MNm/rad/m. These values can be compared with the values of 75, 50 and 400 calculated previously. The higher values here for K_x and K_m are indicative of the relatively stiff paths for horizontal reactions and moments through the ground between the portal's footings.

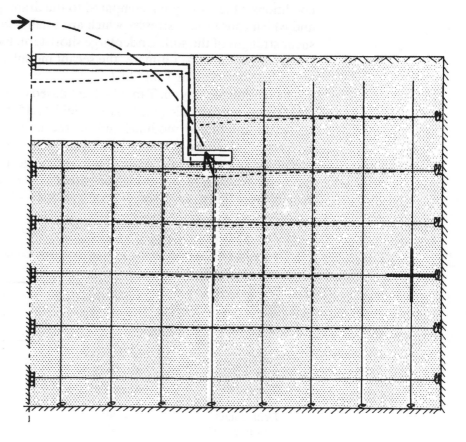

Fig. 14.8 Section of cruciform space frame of Fig. 14.7, with line of thrust across portal under distributed loading.

14.4 STIFFNESS MODULI OF SOILS

The stiffnesses of the ground beneath a bridge should be obtained from the site investigation. This is not straightforward, because tests on small samples seldom provide reliable information on the stiffness of the ground *en masse*. Stiffnesses are best estimated from large-diameter plate bearing tests and from back analyses of observations of comparable structures on similar ground conditions.

Table 14.1 provides an approximate guide to the orders of magnitude of Young's moduli of various soils, based on information in references [1]–[6]. Reference [6] provides more specific guidance and a useful review of other published information. Information in the table may be found helpful for preliminary assessments of structure–soil interaction, but it is not recommended for final design unless corroborated by investigations of the specific site and foundation conditions.

The modulus for clay depends on whether the material has time to drain under the loading and on the amount of strain caused by the loading. The figures in Table 14.1 relate essentially to live loading conditions which are quick compared to the drainage time for the soil, and which cause shear stresses which are small relative to the ultimate shear strength of the soil. Under large short-term loadings which create stresses of about 50% ultimate, the secant modulus may be reduced to about half the value in Table 14.1.

The moduli for clays in Table 14.1 are based on references [4] and [5] which reported the ratio (E_U/S_U) from field data for various clays, where E_U is the undrained modulus and S_U the undrained shear strength. Reference [4] indicates that (E_U/S_U) is in the range 400 to 800 for highly plastic clays and organic clays, and in the range 1000 to 1500 for lean inorganic clays of mean to high sensitivity. Reference [5] reports the results from back analyses of case histories of structures on over-

Table 14.1 Young's moduli of soils under spread footings during short-term low-strain loading

Soil type	Approximate strength	Young's modulus corresponding to strength in column 2
Stiff clay	$S_U = 100$ kPa	40 (–120) MPa
Very stiff clay	$S_U = 200$ kPa	80 (–240) MPa
Loose sand*	$\phi = 30°$	20–60 MPa
Medium sand*	$\phi = 35°$	40–120 MPa
Dense sand*	$\phi = 40°$	80–240 MPa
Dense gravel*	$\phi = 45°$	160–480 MPa

* Under vertical pressure of about 200 kPa.

consolidated clays which correlated well with $(E_U/S_U) = 400$. Table 14.1 shows values of E based on $(E/S_U) = 400$ with values in brackets based on $(E/S_U) = 1200$. The lower values at the bottom of the range are more appropriate to the stiff over-consolidated clays which are suitable for spread foundations for bridges and are widespread in the United Kingdom. Soft clays are not listed in Table 14.1 because spread footings on soft clay are unlikely to be appropriate for bridges, and piles are likely to be adopted. Reference [3] provides guidance on appropriate moduli to be used for assessing the stiffnesses of soils interacting with pile foundations.

The modulus E for a clay under long-term drained conditions is substantially lower than under short-term conditions. Reference [5] reports values of E/S_U of the order of 130 for drained behaviour of over-consolidated clays as compared to 400 for initial undrained deformations. The modulus depends on the previous loading history of the ground, such as if it has been unloaded prior to reloading (by excavation and demolition).

The example stiffnesses for the foundations of the portal bridge in Section 14.3 (and Section 2.7) are based on Young's modulus for the soil of $E = 100$ MPa which relates to a very stiff clay with $S_U = 250$ kPa under undrained live load conditions. In Section 2.7 the same soil is considered under longer term conditions with $E = 100/3$ MPa.

The moduli for sands and gravels depend on the confining pressure and the values in Table 14.1 correspond approximately to a vertical pressure of about 200 kPa. References [1] and [3] explain how E can be assumed to vary in proportion to the pressure, or the square root of the pressure.

Poisson's ratio is generally assumed to be 0.5 for undrained conditions so that the shear modulus $G = E/3$. Under drained conditions Poisson's ratio is about 0.2 for medium and stiff clays, and 0.3 for sands, so that $G = E/2.5$ approximately.

14.5 STIFFNESSES FROM LATERAL EARTH PRESSURES

The distribution of horizontal forces on the supports of a bridge depends on their relative horizontal stiffnesses. The braking forces on the structure of Fig. 14.9(a) may be shared between the lateral flexing of the piers and the passive reactions at an abutment. Thermal expansion of the deck in Fig. 14.9(b) is resisted by flexure of the piers and passive resistance at both abutments. The stiffnesses of abutments under lateral loading are different under passive loading (when the abutment moves towards the embankment) and under active loading (when the abutment moves away from the embankment). The passive stiffness can

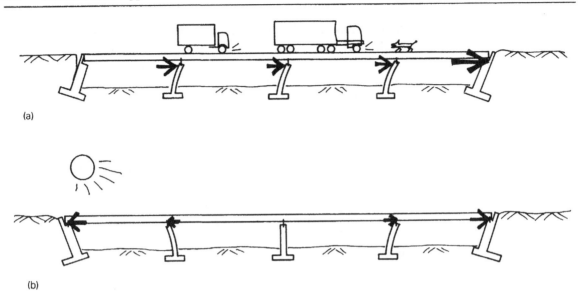

Fig. 14.9 Horizontal forces on bridge: (a) due to braking or traction and (b) due to thermal expansion.

be ten times greater than the active stiffness. The stiffnesses may change after the abutments have been subjected to numerous cycles of thermal movements. All estimates of lateral stiffness should be treated with caution, and great care should be taken if the calculated design load on part of a bridge depends on such a stiffness.

The lateral stiffness of an abutment can be estimated from diagrams of earth pressure versus displacement in texts on foundations, such as reference [7]. References [1] and [7] provide examples of the increase in earth pressure coefficient for granular back-fill with displacement of the top of the wall. The earth pressure coefficient K increases from about 0.4 at rest to about 1.5 when the displacement is 0.002 of the wall height, to about 2.5 at 0.01, and about 4 at 0.03 displacement/height. The displacements of abutments under braking loads and thermal expansion are unlikely to mobilize the full passive resistance, except at the ends of long bridges.

The following calculation demonstrates how the horizontal stiffness can be calculated from earth pressure resistance and rocking rotation of an abutment. (Section 14.7.2 provides an example relating horizontal stiffness to the bodily movement of a small integral abutment.) Consider the abutment of Fig. 14.10 of height $H = 6$ m. Initially guess a displacement of 12 mm at deck level i.e. 0.002 of the height. For this displacement the earth pressure coefficient would change from about 0.4 to 1.5. The earth resistance force per unit width of abutment, with retained fill of density γ, is

$$P = K\gamma H^2/2. \tag{14.9}$$

Hence the change in force, due to K changing from 0.4 to 1.5 with

Fig. 14.10 Abutment resisting horizontal forces from deck.

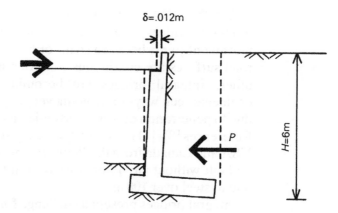

$\gamma = 0.020 \, \text{MN/m}^3$, is

$$P = (1.5 - 0.4) \times 0.020 \times 6^2/2 = 0.40 \, \text{MN/m}.$$

The resultant earth pressure force is assumed to act at one-third height (2 m) up the abutment wall. Thus the resistance provided at deck level is (by taking moments about the foundation)

$$R = 0.40 \times 2/6 = 0.13 \, \text{MN/m}.$$

Since the assumed displacement at deck level is 0.012 mn, the effective stiffness K_x per metre width is

$$K_x = 0.13/0.012 = 11 \, \text{MN/m/m}.$$

An abutment 12 m wide would provide a stiffness of 132 MN/m, that is, a resistance force of 1.6 MN for the displacement of 12 mm.

Active pressure stiffnesses can be analysed in the same manner. However, because active pressures are so much lower than passive forces, it is often sensible to ignore their contribution to stiffness.

14.6 EMBANKMENT MOVEMENTS

The preceding discussion on stiffnesses of foundations has been based on the assumption that the ground is essentially stationary. If the ground is moving due to the construction of embankments and settlement, or due to excavation of a large mass of ground, then careful attention must be paid to the structure–soil interaction as discussed in reference [7]. Very large forces may accompany the movement of a large mass of ground. A bridge can be designed to be stiff, such as a box culvert, or to articulate. If it is stiff it must be designed to be strong enough to resist the forces from the interaction with ground movements. If it articulates it must be designed to accommodate the displacements of the ground movement.

14.7 INTEGRAL BRIDGES

'Integral bridges' are bridges which are constructed without any movement joints between spans or between spans and abutments. The road surfaces are continuous from one approach embankment to the other. Integral bridges are becoming increasingly widespread as engineers seek ways of avoiding very expensive maintenance problems due to penetration of water and de-icing salts through movement joints. References [8]–[11] describe the steady spread of integral bridges in the USA and Canada from the 1930s. It is now not uncommon for bridges to be built without any joints with overall lengths in concrete over 200 m and in steel over 100 m.

Integral bridges present a challenge for load distribution calculations because the bridge deck, piers, abutments, embankments and ground must all be considered as a single compliant system. Not only is the overall system more complicated than a single deck or foundation, but the uncertainties about material stiffnesses are much more significant. The analysis of a simply supported span is not sensitive to an error in material stiffness if the whole structure is affected simultaneously. However with an integral bridge all the material and structural stiffnesses must be estimated as realistically as possible because the load distribution depends on the relative stiffnesses of all the components. Computer methods for load distribution are potentially very useful because of the ease with which one can investigate the possible ranges of behaviour that correspond to the possible ranges of stiffnesses. However the increasing popularity of integral bridges is a result of their successful performance rather than appropriate calculations.

Integral bridges are generally designed with the stiffnesses and flexibilities spread throughout the structure/soil system, so that no part forms a particularly hard spot or a particularly soft spot. The piers and

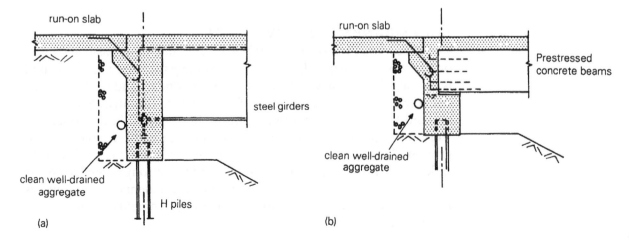

Fig. 14.11 Integral abutments: (a) steel girder composite deck and (b) prestressed beam composite deck.

abutments have sufficient 'give' to absorb thermal expansion and contraction of the deck. At the same time the piers and abutments have sufficient stiffness to resist longitudinal forces due to braking, etc. Examples of integral abutments are shown in Fig. 14.11. The principles of the design are explained in references [8]–[15], which have other examples of details. In general the foundations are designed to be small and flexible to facilitate horizontal movement or rocking of the support. Piles, if used, are in a line so that they can rock. Footings rest on a granular bed that accommodates sliding. Wing walls are small so that they can move with the abutment. Abutments are small in order to limit the amount of passive resistance of the back-fill. Back-fill is not compacted. Run-on-slabs are used to prevent traffic compacting the back-fill disturbed by abutment movements. The residual longitudinal movement between the end of the run-on slab and the pavement is absorbed by yielding of the asphalt pavement, possibly with a plug joint. (Deterioration of the asphalt pavement is found in general to be less expensive to repair than corrosion of substructures.) Concrete pavements need to have compression joints at the ends of the run-on slabs.

The movements of integral abutments do not cause as much damage as might be expected to pavements and embankment fill because the maximum daily changes in effective bridge temperature are very much smaller than the maximum ranges shown in codes between extreme summer maximum and extreme winter minimum [16]. (The effective bridge temperature is the average temperature which governs longitudinal movement.) The daily cycles of movement on many bridges may be accommodated by elastic deformation of pavements and embankments, while the larger seasonal and yearly cycles may be absorbed to a certain extent by creep.

Bridges with rigidly fixed abutments (as opposed to moving integral abutments) may also benefit from the relative smallness of daily ranges of effective temperature and may be designed to accommodate locked-in temperature stresses (as opposed to temperature movements), like continuous railway track. A very long jointless steel viaduct may be designed with fixed abutments so that daily cycles of locked-in stress cause little fatigue as compared to live load, while extreme changes are accommodated by the ultimate strength of the deck and passive resistance of ground at abutments. The size of the fixed abutments here depends on the thrust from the deck cross-section and not the length; so that economy increases with length as compared to a viaduct with movement joints.

The interaction of an integral bridge with its environment can be analysed with a global model, as is demonstrated in Fig. 14.12. (a) shows a longitudinal section of half of the length of the bridge with run-on slab, while (b) shows a plane frame model for the analysis of longitudinal forces and movements due to temperature and braking. The deck is continuous over the piers and is built into the abutments. The run-on slabs are modelled with pin-joints at the abutments. Each abutment is supported by two spring supports and restrained horizontally by lateral springs on the footing and at the centroid of passive earth pressure. Each pier is supported on two spring supports and restrained horizontally by a lateral spring. Each pier here has a moment release at the top to simulate a bearing with dowel.

An integral bridge which has a skew or curvature is likely to require a space frame for the global analysis, in order to take account of the differences in direction of loading and support restraints.

14.7.1 Four-span integral bridge example

The integral bridge in Fig. 14.12 has spans of 20 m and 15 m and stands on footings of 14 m × 3 m under the piers and 14 m × 2 m under the abutments. The spring stiffnesses of the supports are calculated with the equations in Section 14.3. For example, under the action of short term braking forces, the soil might have Young's modulus $E = 40$ MPa. The footings of the piers have $A = 42 \text{ m}^2$ and $Z = 21 \text{ m}^3$ and equations (14.5)

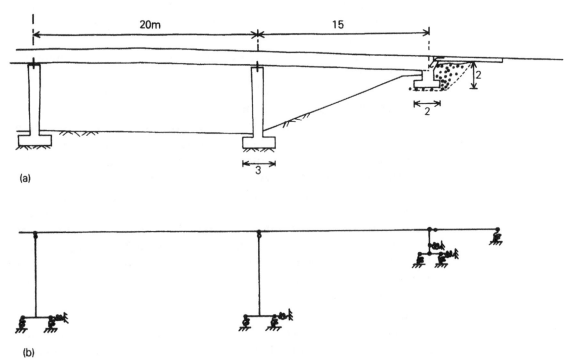

Fig. 14.12 Half of an integral bridge: (a) longitudinal section and (b) plane frame model for global analysis of structure and foundations.

to (14.8) give

$$2 \text{ vertical springs} \quad K = 0.5 \times 1.5EA^{0.5} \quad = 200 \text{ MN/m}$$
$$\text{at spacing} \quad l = 0.82 \times 14^{0.25} 3^{0.75} = 3.6 \text{ m}$$
$$\text{horizontal springs} \quad K_x = EA^{0.5} \quad = 260 \text{ MN/m}.$$

In this example it is assumed that at the abutments the granular bed is medium-dense uniform rounded gravel with $\phi = 35°$, and that the backfill is loose with $\phi = 30°$. (The ϕ angles should be based on tests on typical material; here they are taken from reference [1].) The Young's modulus of the soil under the abutment granular bed is about 40 MPa. The footings have $A = 28 \text{ m}^2$ and $Z = 9 \text{ m}^3$ and equations (14.5) to (14.8) give

$$2 \text{ vertical springs} \quad K = 0.5 \times 1.5EA^{0.5} \quad = 160 \text{ MN/m}$$
$$\text{at spacing} \quad l = 0.82 \times 14^{0.25} 2^{0.75} = 2.7 \text{ m}$$
$$\text{horizontal springs} \quad K_x = EA^{0.5} \quad = 210 \text{ MN/m}.$$

The lateral restraints to the abutments are non-linear, and equivalent springs have to take account of the soil mechanics. An abutment being pulled away from the embankment is subject to active pressures, and its contribution to stiffness may be ignored. An abutment being pushed into the embankment is resisted by sliding and passive pressure. In this case the small abutment is fixed to the deck and the abutment is assumed to move bodily rather than rotate.

There is an upper limit to the horizontal resistance when sliding occurs. If the vertical reaction is 3 MN the sliding friction is

$$F = W \tan \phi = 3 \tan 35° = 2.1 \text{ MN}$$

which occurs after a displacement $x = F/K_x = 2.1/210 = 0.01 \text{ m}$. It is found under ultimate temperature loads that the displacement just exceeds 0.01 m and then the horizontal springs on the abutment footings are replaced by the limiting forces $F = 2.1$ MN resisting sliding.

The passive resistance spring needs to be based on test results of passive resistance of granular fill under cyclic loading. If the retained fill is only 2 m high and it is displaced 0.01 m, this displacement represents 0.005 of the height. The strain in the back-fill will be of the same order, and reference [1] indicates that at this strain the earth pressure is only half the fully mobilized passive value. Hence K is here about 1.5, in which case the resisting force from the 14 m width of back-fill of $\gamma = 0.016 \text{ MN/m}^3$ is, from equation (14.8),

$$P = 1.5 \times 0.016 \times \frac{2^2}{2} \times 14 = 0.67 \text{ MN}$$

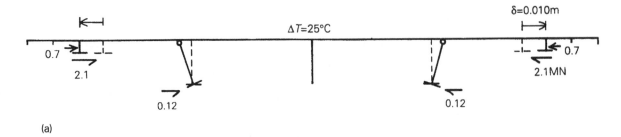

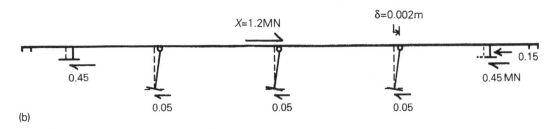

Fig. 14.13 Distributions of reactions under integral bridge: (a) due to temperature rise of 25°C and (b) due to braking force of 1.2 MN.

and the equivalent spring has

$$K_x = 0.67/0.01 = 67\,\text{MN/m}.$$

Figure 14.13 shows the distributions of displacements and horizontal reactions calculated for loadings: (a) a temperature increase of 25°C, and (b) a braking force of 1.2 MN. By using two vertical springs under each footing it is easy to check that no spring goes into tension under combined weight and temperature loading. A tensile reaction would be indicative of uplift, in which case the foundation model would have to be modified to represent the reduced area of contact in compression, or the design modified to make the structure or foundation more flexible.

This example has considered a change in effective bridge temperature of 25° relative to the mean. This large temperature change represents half an ultimate range from summer maximum to winter minimum over a return period of very many years. The maximum range for any one day may be only about one fifth of the ultimate range. On this short bridge the maximum daily range of movements of the abutments would then be only about 4 mm, i.e. +2 mm to −2 mm relative to the mean, and the responses of the ground and pavement at each end would be largely elastic. However the movements at the ends of a long bridge may well exceed the elastic range for the ground and embankment, and the analysis will need to consider yield and/or sliding as in this example. A rigorous analysis of the movements over a season, or a few years, is more complicated and needs to consider the creep relaxation of the ground, pavement and structure, which can eliminate much of the resistance force.

Fig. 14.14 Composite prestressed concrete deck of Long Island Bridge, Kingsport, Tennessee; designed by Tennessee Department of Transportation. This 850 m long jointless bridge deck has expansion joints only at the abutments.
Photograph courtesy of George Hornel, State of Tennessee.

The calculation can be carried out for upper bound and lower bound estimates of soil strength and stiffness.

The analysis of settlements of the supports and associated deck moments may need a detailed comparison of the settlement characteristics of the different soils, and of the embankments. Much of the differential settlement may occur during construction before the deck is made continuous, as explained in reference [7].

REFERENCES

1. Lambe, T.W. and Whitman, R.V. (1969) *Soil Mechanics*, John Wiley, New York.
2. Richart, F.E., Hall, J.R. and Woods, R.D. (1970) *Vibrations of Soils and Foundations*, Prentice-Hall, New Jersey.

3. Poulos, H.G. and Davis, E.H. (1980) *Pile Foundation Analysis and Design*, John Wiley, New York.
4. D'Appolonia, D.J., Poulos, H.G. and Ladd, C.C. (1971) Initial settlements of structures in clay, *Proceedings of American Society of Civil Engineers, Journal of Soil Mechanics and Foundation Engineering Division*, **97**, SM10.
5. Butler, F.G. (1974) 'Heavily over-consolidated clays', Settlement of Structures, Conference of British Geotechnical Society, Pentech Press, London, 1975.
6. Wroth, C.P., Randolph, M.F., Houlsby, G.T. and Fakey, M. (1984) 'A review of the engineering properties of soils with particular reference to shear modulus', Oxford University Engineering Laboratory Report 1523/84.
7. Hambly, E.C. and Burland, J.B. (1979) *Bridge Foundations and Substructures*, Her Majesty's Stationery Office, London.
8. Federal Highway Administration (1980) Technical Advisory T5140.13, 'Integral, no-joint structures and required provisions of movement', US Department of Transportation, Washington DC.
9. Loveall, C.L. (1985) Jointless bridge decks, *Civil Engineering, American Society of Civil Engineers*, New York, NY, November.
10. Wasserman, E.P. (1987) Jointless bridge decks, *Engineering Journal, American Institute of Steel Construction*, Chicago, 3rd quarter.
11. National Cooperative Highway Research Program, NCHRP 141 (1989) 'Bridge deck joints', by Martin P. Burke Jr, Transportation Research Board, Washington DC.
12. National Cooperative Highway Research Program, NCHRP 322 (1989) 'Design of precast prestressed concrete bridge girders made continuous', by R.G. Oesterle, J.D. Glikin and S.C. Larson, Transportation Research Board, Washington DC.
13. Burke, M.P. (1990) 'The integrated construction and conversion of single and multiple span bridges', in *Bridge Management* (eds J.E. Harding, G.H.R. Parke and M.J. Ryall) University of Surrey 1st International Conference on Bridge Management, Elsevier Applied Science.
14. Hambly, E.C. and Nicholson, B.A. (1990) Prestressed beam integral bridges, *Structural Engineer*, **68**, December.
15. Hambly, E.C. and Nicholson, B.A. (1991) *Prestressed Beam Integral Bridges*, Prestressed Concrete Association, Leicester.
16. Emerson, M. (1976) Bridge temperatures estimated from the shade temperature, Laboratory Report 696, Transport and Road Research Laboratory, Crowthorne.

Appendix A

Product integrals

Functions of load

on a single span

Harmonic components

Fig. A.1 Product integrals.

m_1	m_2	$\int m_1 m_2 \, dx$
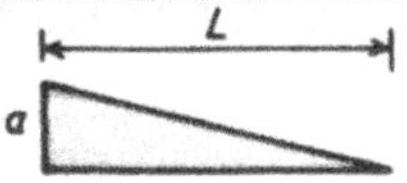 triangle, height a at left, length L	 triangle, height b at left	$\frac{L}{3} ab$
 triangle, height a at left	 triangle, height b at right	$\frac{L}{6} ab$
 triangle, height a at left	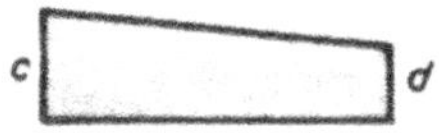 trapezoid, c left, d right	$\frac{L}{6} a(2c+d)$
 triangle, height a at left	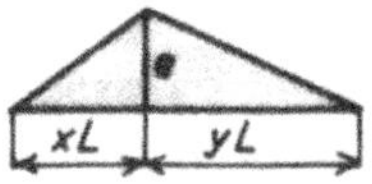 triangle peak e at xL, yL	$\frac{L}{6} ae(1+y)$
 triangle, height a at left	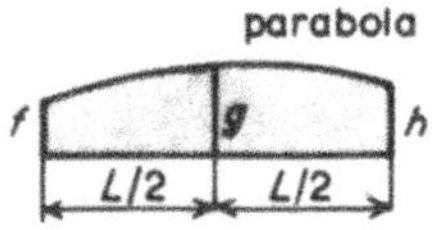 parabola, f, g, h, $L/2$, $L/2$	$\frac{L}{6} a(f+2g)$
 trapezoid, c left, d right	trapezoid, j left, k right	$\frac{L}{6}[c(2j+k)+d(2k+j)]$
 trapezoid, c left, d right	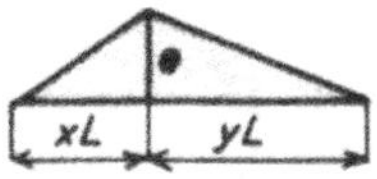 triangle peak e at xL, yL	$\frac{L}{6} e[c(1+y)+d(1+x)]$
 trapezoid, c left, d right	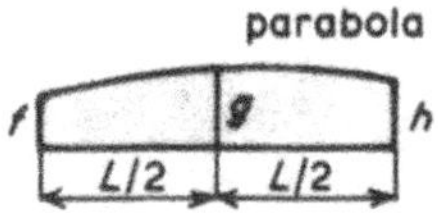 parabola, f, g, h, $L/2$, $L/2$	$\frac{L}{6}[c(f+2g)+d(h+2g)]$
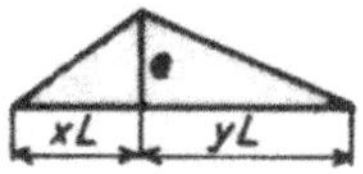 triangle peak e at xL, yL	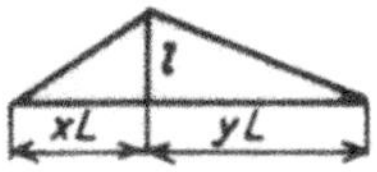 triangle peak l at xL, yL	$\frac{L}{3} el$
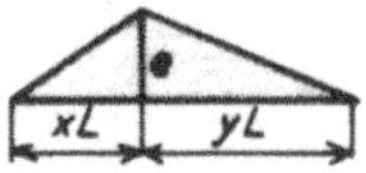 triangle peak e at xL, yL	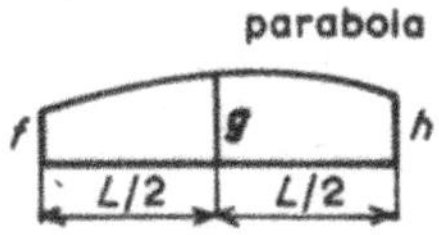 parabola, f, g, h, $L/2$, $L/2$	$\frac{L}{6} e[fy^2 + hx^2 + 2g(1+xy)]$

Fig. A.2 Functions of loads on a simple span.

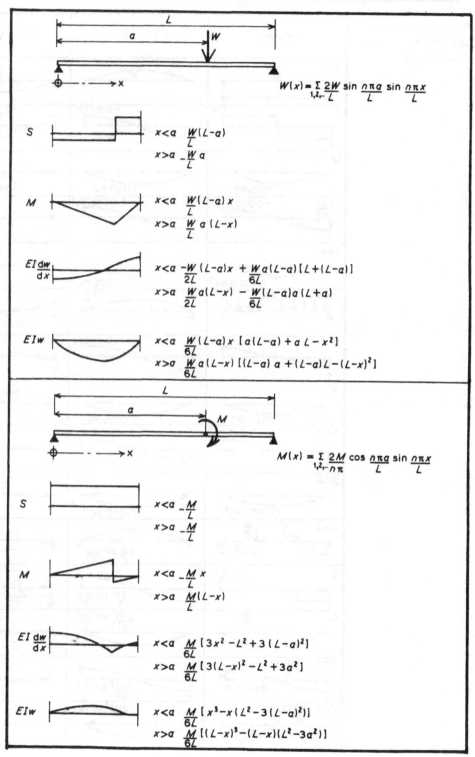

Appendix A

Fig. A.3 Harmonic composition of loads, moments, etc.

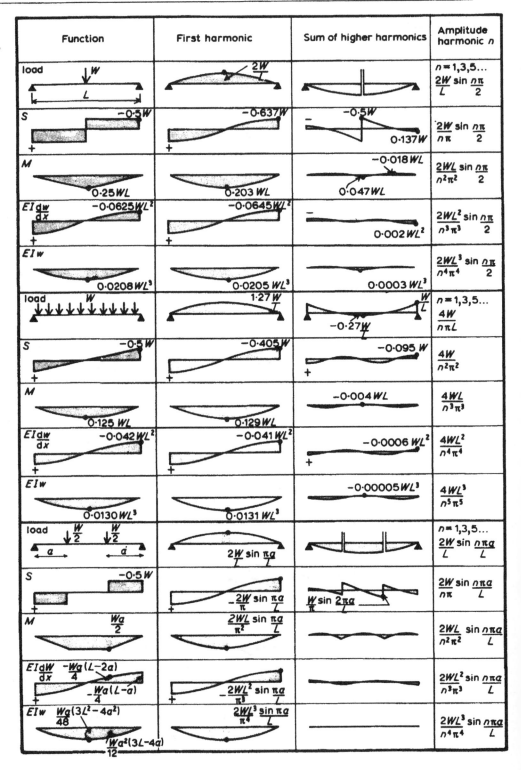

Appendix B Calculation of torsion constant for solid beams

B.1 MEMBRANE ANALOGY BY FINITE DIFFERENCE METHOD

The torsion constants of solid beams of complex section can be calculated by using Prandtl's membrane analogy and the finite difference method. The method is described in detail in reference [1] and summarized below. The procedure is relatively simple and can be executed quickly by a program on a microcomputer.

The cross-section of the beam is subdivided as in Fig. B.1 into a fine grid. The fineness of the grid is determined by the narrowness of the neck where shear stresses are likely to be highest. Reference [1] explains how the method can be made to fit curved boundaries; here it is assumed that the stepped boundary in Fig. B.1 is adequate. The membrane

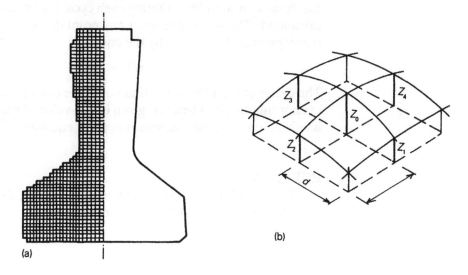

Fig. B.1 Membrane analogy for torsion of beam: (a) section and calculation grid and (b) local geometry of membrane.

analogy takes advantage of the similarity of the equations for the stress function for a section under torsion and the equations for the shape of an inflated soap bubble stretched over a boundary of the same shape. The shape of the soap bubble is controlled by the equation

$$\frac{\partial^2 z}{\partial x^2} + \frac{\partial^2 z}{\partial y^2} = \frac{p}{t} \qquad (B.1)$$

where z is the elevation of the bubble as shown in Fig. B.1(b), p is the pressure under the bubble and t is the surface tension.

The torsion constant C is related to the volume v of the soap bubble by the equation

$$C = \frac{4t}{p} \cdot v. \qquad (B.2)$$

The volume of the bubble for any value of t/p can be found by solving equation (B.1) at each point of the mesh using successive approximations of the finite difference equation

$$z_0 = \frac{1}{4}\left(\frac{pd^2}{t} + z_1 + z_2 + z_3 + z_4\right) \qquad (B.3)$$

where z_0, z_1, z_2, z_3 and z_4 are the elevations at points 0, 1, 2, 3 and 4, respectively, and d is the grid spacing, as shown in Fig. B.1(b). By arbitrarily setting $p = 4t$ equation (B.2) makes $C = v$ and equation (B.3) is simplified to

$$z_0 = d^2 + (z_1 + z_2 + z_3 + z_4)/4. \qquad (B.4)$$

It is assumed initially that values of z are all zero over the grid. Then by solving equation (B.4) for each point in turn, values of z are established. By working over the grid several times the elevations gradually take up the shape of a soap film. During each cycle the volume under the film is calculated. The shear stress τ at any point due to torque T is related to the slope m and volume v by the equation

$$\tau/T = m/2v. \qquad (B.5)$$

The cycles of calculation continue until the changes in v and m per cycle are not significant. Then C is given by the value of v and maximum (τ/T) is calculated from the maximum m using equation (B.5).

REFERENCE
1. Timoshenko, S.P. and Goodier, J.N. (1970), *Theory of Elasticity*, McGraw-Hill, New York, 3rd edn.

Author index

Abdel-Samad, S.R. 135–41, 156
Abel, J.F. 263, 280
Ahmad, S. 263, 280
American Association of State Highway and Transportation Officials (AASHTO) 180, 187, 193, 198
American Institute of Steel Construction 59, 81
Armer, G.S.T. 192, 198

Bakht, B. 61, 81, 199, 221, 253, 261, 262
Balas, J. 191, 198, 200, 221
Beckett, D. 1, 23
Best, B.C. 75, 81
British Standard 137, 156, 164, 176, 180, 187
Buckley, R.J. 91, 99, 100, 105
Burke, M.P. 294, 295, 300
Burland, J.B. 292, 293, 299, 300
Burnett, D.S. 263, 280
Butler, F.G. 290, 291, 300

Comité Euro-International du Béton 50, 52, 89, 104
Cheung, Y.K. 263, 276, 280
Clark, L.A. 50, 51, 52, 90, 105, 192, 198, 227, 242
Clough, R.W. 263, 280
Coates, R.C. 38, 52
Cohen, E. 50, 51, 52
Coutie, M.G. 38, 52
Cusens, A.R. 201, 221, 253, 262, 263, 276, 280

D'Appolonia, D.J. 290, 300

Davis, E.H. 286, 290, 291, 300
DeFries-Skene, A. 255, 256, 262
Department of Transport 201, 221
Desai, C.S. 263, 280

Emerson, M. 223, 242, 296, 300
Evans, R.H. 50, 51, 52, 89, 104
Evans, H.R. 263, 270, 280

Fakey, M. 290, 300
Federal Highway Administration 294, 295, 300

Gere, J.M. 28, 31, 40, 42, 52
Goldberg, J.E. 255, 256, 262
Goodier, J.N. 42, 43, 45, 52, 305, 306
Griffiths, D.W. 263, 270, 280

Hall, J.R. 283, 299
Hambly, E.C. 90, 104, 106, 132, 134, 205, 206, 221, 223, 227, 242, 258, 262, 292, 293, 295, 299, 300
Hanuska, A. 191, 198, 200, 221
Harris, J.D. 231, 243
Hayward, A.C.G. 91, 99, 100, 105
Hendry, A.W. 253, 262
Hergenroder, A. 191, 198, 200, 221
Hooke, R. 7, 23
Horne, M.R. 5, 23
Houlsby, G.T. 290, 300

Iles, D.C. 91, 99, 100, 105
Irons, B. 263, 280

Jaeger, L.G. 61, 81, 199, 221, 253, 261, 262

Author index

Johnson, R.P. 91, 99, 100, 105

Kong, F.K. 38, 50, 51, 52, 89, 104
Kreyszig, E. 245, 253, 262

Ladd, C.C. 290, 300
Lambe, T.W. 283, 290, 291, 292, 297, 299
Lee, D.J. 1, 23, 137, 156
Leonhardt, F. 1, 23
Leve, H.L. 255, 256, 262
Libby, J.R. 231, 243
Liebenberg, A.C. 1, 23
Lightfoot, E. 32, 33, 52, 61, 81
Little, G. 200, 208, 212, 221
Loo, Y.C. 263, 276, 280
Loveall, C.L. 294, 295, 300

Maisel, B.I. 137, 156
Martin, H.C. 263, 280
McHenry, D. 160, 176
Morice, P.B. 36, 52, 200, 208, 212, 221

Nakai, H. 136, 156, 180, 187, 193, 195, 198
National Cooperative Highway Research Program 294, 295, 300
Nethercot, D.A. 263, 270, 280
Nicholson, B.A. 90, 104, 227, 242, 295, 300

Oden, J.T. 28, 40, 42, 52
Ontario 180, 187

Pama, R.P. 201, 221, 253, 262
Pennells, E. 90, 93, 104, 106, 134
Perkins, N.D. 231, 243
Poulos, H.G. 286, 290, 296, 300
Prestressed Concrete Institute 137, 156
Priestley, M.N.J. 223, 242
Pucher, A. 61, 74, 81, 93, 105, 127, 134, 200, 221

Randolph, M.F. 290, 300
Richart, F.E. 283, 290, 299
Richmond, B. 1, 23, 137, 156
Roark, R.J. 28, 40, 42, 52, 180, 187
Robinson, A.R. 135–41, 156
Rockey, K.C. 263, 270, 280
Roll, F. 50, 51, 52, 137, 156
Rowe, R.E. 1, 81, 93, 105, 200, 221, 253, 262
Rusch, H. 191, 198, 200, 221

Sawko, F. 61, 81, 106, 134, 220, 221
Scordelis, A.C. 255, 256, 262
Shrive, N. 263, 280
Smith, I.C. 231, 243
Spindel, J.E. 75, 81
St Venant, B. 6, 23
Steel Construction Institute 227, 242

Taylor, R.L. 263, 280
Timoshenko, S.P. 28, 31, 40, 42, 43, 45, 52, 61, 81, 305, 306
Topp, L.J. 263, 280
Troitsky, M.S. 59, 81
Turner, M.J. 263, 280

Wasserman, E.P. 294, 295, 300
West, R. 63, 81
Whitman, R.V. 283, 290, 291, 292, 297, 299
Woinowsky-Krieger, S. 61, 81
Wood, R.H. 192, 198
Woods, R.D. 283, 290, 299
Wright, R.N. 137–41, 156
Wroth, C.P. 290, 300

Yoo, C.H. 136, 156, 180, 187, 193, 195, 198
Young, T. 22, 23
Young, W.C. 28, 40, 42, 52, 180, 187

Zienkiewicz, O.C. 263, 280

Subject index

AASHTO 96, 187, 198
Abutments 291–9
Active stiffness 292
Antisymmetry *see* Symmetry and antisymmetry
Approximations and errors
   accuracy xv
   loading xvi
   material behaviour xvi, 17, 285, 290, 294
   mistakes xv, 18, 90
   numerical method 21, 160, 261, 268
   physical simulation xvi, 52, 95, 104, 115, 145, 212, 279
   structural theory 43, 61, 180, 257
Arching action 47–50, 285
Assessment xiii, xiv
Axial stress 47, 172

Back fill 292, 295
Battledeck 11, 67
Beam-and-plank deck 253
Beam-and-slab decks 8, 82–105, 167–76, 191, 201, 205, 209, 213–16
Beam decks 2, 24–52
Beam element 274
Beam-on-elastic-foundations 136–54
Bearings 64, 96, 136, 189, 281
BEF method *see* Beam-on-elastic-foundations
Bending inertia *see* Section properties *and* Grillage member properties
Bending of beam 26–39, 84, 194

Bending of cellular deck 108–13
Bending of slab 55–60
Bending stress 28, 108, 111, 181
Box beams 7, 44, 78, 82, 91, 95, 190, 213
Box culvert 293
Box girders (*see also* Cellular decks) 12–15, 29, 45, 106, 116, 120, 122, 135–56, 164, 240, 269, 282
Bracing 99–101, 136–53
Braking forces 291, 296
Britannia Bridge 2, 46
Brown & Root 176
Buckling 18, 157, 180
Butterley Engineering Ltd 220

Cantilever construction 15
Cass Hayward & Partners 103
Cellular decks 12, 106–34, 201, 205, 230 (*see also* Multicellular decks)
Cellular stiffness ratio 205
Charts 191, 199
Chichester 220
Clays 290
Collapse 182 (*see* also Ultimate limit state)
Complementary 29, 56, 62
Composite construction 7, 53, 69–71, 89, 91, 96, 97, 137, 157, 169
Computer methods xiv, xv, 1, 45, 53, 137, 263, 279, 294
Codes of practice xiii, xiv
Concentrated loads 63–4, 73, 93, 179, 200, 201, 251

Subject index

Conforming elements 273, 276
Construction sequence 45–7, 50, 184
Contiguous beam-and-slab 10, 83, 86, 89, 174–5, 216
Continuous decks 24, 25, 181, 206, 245, 251, 257, 260, 282
Contraflexure *see* Continuous decks
Cracked concrete 70, 80, 89, 90, 158, 227
Creep 20, 50, 51, 230, 239, 295, 298
Cruciform space frame 142, 159, 161–5, 239, 288
Curved decks 136, 159, 193–8, 240–1, 296

Deflections 30
Design xiii, xiv, 17–21
Detailed analysis 19
Determinate structures 20, 24–6, 40
Diaphragms 3, 8, 15, 83, 86, 90, 95, 106, 113, 120, 136, 157, 206, 256
Differential settlement 119, 136, 281
Dispersal 73
Displacement field 265, 272
Distortion
  box 17, 92, 135–56
  multicell 15, 106, 115–20, 129–31, 201
  slab in-plane 167
Distribution coefficients 97, 199–221
Downstand grillage 167–76
Drained soil 49, 291
Ductility 18

Earth displacement 292
Earth pressures 291–2
East Moors Viaduct 13
Economy of calculation xv
Edge beams *see* Edge stiffening
Edge stiffening 54, 71, 182–5, 192
Effective bridge temperature 295, 298
Effective flange width 88, 98, 164, 175, 177–82
Effective modulus 51
Elastic half space 283
Elastic loss *see* Prestress losses

Elastic theory 17–22
Embankments 294
Embankment movements 293
Equilibrium (*see also* Statical equivalence) 17–21, 268, 270–1

Fatigue 19
Festival Park Flyover 103
Finite difference method 305
Finite elements 8, 17, 21, 137, 141–7, 159, 221, 263–80, 286
Finite strip analysis 132, 276
Fixed end forces 32
Flexibility methods 35–8, 236, 271
Flexural rigidity 56–8, 77, 195
Flexural stiffness ratio 205
Folded plate theory 17, 183–5, 205, 255–61
Foundations 49, 281–99
Foundation stiffnesses 283–99
Fourier analysis 245
Frame action *see* Arching action
Framing 136, 144, 151, 157
Friction loss *see* Prestress losses

Gibb's phenomenon 261, 277
Gifford & Partners ii, xi, 121, 131, 279
Global analysis 19, 93, 182, 296
Grampian Regional Council 9
Greta Bridge 242
Grid decks 3
Grillage
  beam-and-slab 85–99
  cellular 106–33, 145–54, 278
  local 93–4, 127
  member properties 65–72, 87–92, 108–28, 145–54, 181
  merits xv, 20
  mesh 63–5, 85–7, 107–8, 168, 191–3, 197
  method 61–81, 85–99
  physical simulation 4, 53, 102–4, 184, 186, 275
  skew 78, 89–99, 122–5, 191
  slab 61–81
Groot Lemmer BV 262
G. Maunsell & Partners 80

Hand methods xv, 137, 199
Harmonic analysis 200, 244–53, 277, 304

Subject index

Haunched deck 228, 234, 236
Honshu Shikoku Bridge Authority 14
Hooke, Robert 21
Hooke's Law 21, 224
Houston 176

Inclined webs 107, 116, 125–7
Indeterminate structures 20, 24–6, 40
Influence lines 200–8
Integral bridges 12, 294–9
In-plane behaviour 102–4, 162, 166, 256, 264, see also Membrane action
Isotropic slab 5, 54, 200, 209

Jointless viaduct 295

Kings Langley 155
Kocher Viaduct 4
Kylesku Bridge 197

Launching 16
Line of thrust 49, 289
Load application 18–9, 92–4, 127
Local analysis 19, 64, 92–4, 182
Locked-in stresses 224, 295
Long term 20, 49–51, 290–1
Lower bounds 49, 271, 274, 299
Lower bound theorem 18

M42 121
Maintenance 20, 24, 294
McHenry lattice 159, 160
Meganewton (MN) xx
Megapascal (MPa) xx
Member spacing see Grillage mesh
Membrane analogy 43–44, 305
Millbrook flyover ii
Models 3, 117–8, 135, 178, 220, 232
Modular ratio 70, 90, 91, 98
Mohr's circle 59–60
Moments see Bending
Moment curvature equations 31, 56, 66
Moment of inertia see Section properties
Movement joints 20, 24, 294
Multicellular deck 124, 133, 232 (see also Cellular decks)
Mutliple box girder 149–51, 153

Multispan box girder 151–4

Nashville 12
Neutral axis 28, 103, 110–3, 174–5, 183–6
Newburgh 9
Nominal members 92–3, 107, 116, 122–3, 148, 157
Non-conforming see Conforming elements
Non-linear 17, 178, 297
Notation xvii–xix

Ontario Ministry of Transportation xi, 187
Oosterschelde 16
Orthotropic slabs 7, 54, 58, 67, 200
Orthotropic plate theory 58, 244
Output interpretation 72–3, 94–5, 127–32, 157, 171–3
Out-of-plane behaviour see Bending and In-plane behaviour
Ove Arup & Partners xi, 7, 197

Parapets 54, 174, 183–5
Parasitic moments 238–9
Passive stiffness 291–3
Piers 49–51, 282–3, 294
Pile foundation 286, 291, 295
Plane frame 11, 47–51, 78, 92, 100–101, 116, 152–4, 206, 296
Plane stress elements 264–71
Plastic theory 17
Plate bending elements 271–5
Plate elements 275
Point loads see Concentrated loads
Poisson's ratio 41, 57, 113, 160–1, 229, 275, 291
Portal bridge 48, 286
Post tensioning see Prestress
Potential energy 268
Prandtl see Membrane analogy
Prestressed concrete
  cable curvature 234–8
  cable eccentricity 233–4
  losses 240–1
  post-tensioned decks 121, 125–6, 145, 233–4
  pretensioned beams 69, 89, 90, 96, 173–4, 183–6, 239
  tendons 241

Principal moments 59–60
Principal stresses 60–1
Product integrals 37, 237, 302

Quadrilateral elements 270, 273

Raking piers 50
Redbridge Flyover 279
Redistribution 18, 19, 49, 99, 182
Reinforced concrete 68–72, 89, 157, 191–2, 227
Residual stresss 226
Rigorous analysis *see* Approximations
Robert Benaim & Associates 13
Rotational stiffness ratio 205
Run-on-slabs 295–6

Safety of methods 17–21
Safe design theorem 18–21
St Venant's principle 19, 251
St Venant torsion constant *see* Torsion constant
Sarum Hardwood Structures Ltd 262
Saw tooth diagram 73, 128
Scott Wilson Kirkpatrick & Partners 242
Secondary moments 238–9
Second moment of area *see* Section properties
Section properties 28, 31, 42–5, 181 *see also* Grillage member properties *and* Torsion constant
Serviceability 19, 182
Service bays 185–7
Settlement 20, 119, 136, 281, 299
Shear *see* Bending *and* Torsion
Shear area 31, 77–8, 115–20, 148, 152, 161–5
Shear flexibility *see* Shear area
Shear flexible grillage 16, 106–34, 151–4
Shear flows 58, 68, 111–13, 131, 166–74, 179–84
Shear-key decks 8, 74–81, 92, 201
Shear lag 88, 91, 97–9, 112, 123, 163–4, 170, 175, 177–82, 233
Shear modulus 31, 41, 57, 284
Shear stress 28, 43, 45, 109, 226
Sheffield 6

Shell elements 275
Short term 20, 49–51, 290
Shrinkage 20, 184, 230, 239
Sign convention 26, 31, 48, 60
Simple methods 20
Sine wave *see* Harmonic analysis
Skew decks 40, 64, 70, 83, 90, 96–9, 122–5, 133, 188–92, 200, 230, 260, 281, 296
Slab decks 5, 53–81, 189, 208–12
Slab membrane action 102–3, 165–76
Sliding 297
Slope deflection equations *see* Moment curvature equations
Soil moduli 290–1
Soil stiffness 49–50, 283–99
Solid elements 278
Somerset Sub Unit of South West Road Construction Unit 51
Southampton ii, 133, 279
Spaced beam-and-slab 10, 83, 91–103, 165–74, 183–6
Space frame 20, 141–54, 157–76, 195, 238–41
Staples Corner Flyover 232
Statically determinate *see* Determinate structures
Statical distribution *see* Statical equivalence
Statical equivalence 17–21, 63, 68, 93, 108, 127, 252
Steel beams and girders 91, 97, 157–9, 169
Steel box girders 120, 137–55, 180
Stephenson, Robert 2, 46
Stiffeners 157
Stiffnesses 19, 20, 61, 284, 294
Stiffness method 31–5, 38, 256–60, 264–78
Strain energy 36, 268
Structural forms 2–17
Substructures 281–99
Supports 281–99
Symmetry and antisymmetry 100, 135–7, 142, 247–8, 289

Tapered decks 32, 192–3, 234
Taunton 51
Taylor Woodrow Construction 232
Temperature differential 222–30, 241

Temperature effects 222–30, 291, 295, 298
Temple 262
Tennessee Department of Transportation 12, 299
Theory xv
Thermal *see* Temperature
Three dimensional elements 278
Tony Gee and Partners 155
Toronto 187
Torsion
  beam 39–45
  box 45, 131, 135–6, 153–4
  cellular 113–5, 118–9, 131
  constant 42–4, 66–8, 79, 114
  curved deck 159, 195–8, 241
  grillage 66–8, 88, 97–9, 113, 146–8
  stiffness 119, 186, 195, 282
  shear stress 45, 57, 131
  slab 57–9, 79, 189
  warping 100
Torsionless design 95–99
Torsionless grillage 97
Torsion grillage 97, 99
Transfer 240
Transverse bending 88, 93, 112–3, 129–31
Triangular elements 264, 272
Truss space frame 157–60

Ultimate limit state 18–9, 158, 182
Ultimate strength 19
Undrained shear strength 290
Units xx
Uplift 189, 298
Upper bounds 49, 271, 274, 299
Upstand 54, 183, 185

Van Hattum en Blankenvoort NV 16
Voided slab decks 7, 15, 53, 66, 71, 117

Warping 102, 117–9, 169
Wayss and Frytag 4
Westway 80
Working stress *see* Serviceability
W. V. Amsterdamsche Ballast Maatschappij 16

Yield 18, 182
Yokogawa Construction Company 14
Yoshima Loop Bridge 14
Young, Thomas 22
Young's modulus 22, 28
  concrete 49, 90, 98
  effective 51
  soils 49–50, 290–1, 296–7
  steel 98, 138

For Product Safety Concerns and Information please contact our EU representative GPSR@taylorandfrancis.com Taylor & Francis Verlag GmbH, Kaufingerstraße 24, 80331 München, Germany

Printed and bound by CPI Group (UK) Ltd, Croydon, CR0 4YY

08/06/2025

01897007-0016